NATIONAL ACADEMIES *Sciences Engineering Medicine*

NATIONAL ACADEMIES PRESS
Washington, DC

Principles and Practices for a Federal Statistical Agency

Eighth Edition

Melissa Chiu and Jennifer Park,
Editors

Committee on National Statistics

Division of Behavioral and
Social Sciences and Education

Consensus Study Report

NATIONAL ACADEMIES PRESS 500 Fifth Street, NW Washington, DC 20001

This activity was supported by Grant No. SES- 2217307 between the National Academy of Sciences and the National Science Foundation, which provides support for the work of the Committee on National Statistics from a consortium of federal agencies. Also supporting the Committee's work are a cooperative agreement with the U.S. Department of Agriculture and individual agreements with agencies in the U.S. Departments of Commerce, Health and Human Services, Housing and Urban Development, and the Treasury. Additional support was provided by the National Academy of Sciences W. K. Kellogg Foundation Fund. Any opinions, findings, conclusions, or recommendations expressed in this publication do not necessarily reflect the views of any organization or agency that provided support for the project.

International Standard Book Number-13: 978-0-309-72543-9
International Standard Book Number-10: 0-309-72543-7
Digital Object Identifier: https://doi.org/10.17226/27934
Library of Congress Control Number: 2024951703

This publication is available from the National Academies Press, 500 Fifth Street, NW, Keck 360, Washington, DC 20001; (800) 624-6242; http://www.nap.edu.

Copyright 2025 by the National Academy of Sciences. National Academies of Sciences, Engineering, and Medicine and National Academies Press and the graphical logos for each are all trademarks of the National Academy of Sciences. All rights reserved.

Printed in the United States of America.

Suggested citation: National Academies of Sciences, Engineering, and Medicine. (2025). *Principles and Practices for a Federal Statistical Agency, Eighth Edition*. Washington, DC: National Academies Press. https://doi.org/10.17226/27934.

The **National Academy of Sciences** was established in 1863 by an Act of Congress, signed by President Lincoln, as a private, nongovernmental institution to advise the nation on issues related to science and technology. Members are elected by their peers for outstanding contributions to research. Dr. Marcia McNutt is president.

The **National Academy of Engineering** was established in 1964 under the charter of the National Academy of Sciences to bring the practices of engineering to advising the nation. Members are elected by their peers for extraordinary contributions to engineering. Dr. John L. Anderson is president.

The **National Academy of Medicine** (formerly the Institute of Medicine) was established in 1970 under the charter of the National Academy of Sciences to advise the nation on medical and health issues. Members are elected by their peers for distinguished contributions to medicine and health. Dr. Victor J. Dzau is president.

The three Academies work together as the **National Academies of Sciences, Engineering, and Medicine** to provide independent, objective analysis and advice to the nation and conduct other activities to solve complex problems and inform public policy decisions. The National Academies also encourage education and research, recognize outstanding contributions to knowledge, and increase public understanding in matters of science, engineering, and medicine.

Learn more about the National Academies of Sciences, Engineering, and Medicine at **www.nationalacademies.org**.

Consensus Study Reports published by the National Academies of Sciences, Engineering, and Medicine document the evidence-based consensus on the study's statement of task by an authoring committee of experts. Reports typically include findings, conclusions, and recommendations based on information gathered by the committee and the committee's deliberations. Each report has been subjected to a rigorous and independent peer-review process and it represents the position of the National Academies on the statement of task.

Proceedings published by the National Academies of Sciences, Engineering, and Medicine chronicle the presentations and discussions at a workshop, symposium, or other event convened by the National Academies. The statements and opinions contained in proceedings are those of the participants and are not endorsed by other participants, the planning committee, or the National Academies.

Rapid Expert Consultations published by the National Academies of Sciences, Engineering, and Medicine are authored by subject-matter experts on narrowly focused topics that can be supported by a body of evidence. The discussions contained in rapid expert consultations are considered those of the authors and do not contain policy recommendations. Rapid expert consultations are reviewed by the institution before release.

For information about other products and activities of the National Academies, please visit www.nationalacademies.org/about/whatwedo.

COMMITTEE ON THE EIGHTH EDITION OF PRINCIPLES AND PRACTICES FOR A STATISTICAL AGENCY

KATHARINE G. ABRAHAM (*Chair*), Distinguished University Professor, University of Maryland, College Park
MICK P. COUPER, Research Professor, University of Michigan
WILLIAM A. DARITY, JR., Samuel DuBois Cook Professor of Public Policy, African and African American Studies, Economics, and Business, Duke University
DIANA FARRELL, Independent Director and Trustee (at various institutions, including the Urban Institute and the National Bureau of Economic Research)
ROBERT M. GOERGE, Senior Fellow, NORC of the University of Chicago
ERICA L. GROSHEN, Senior Economic Advisor, Cornell University
DANIEL E. HO, William Benjamin Scott and Luna M. Scott Professor of Law, Professor of Political Science and Computer Science (by courtesy), Stanford University
HILARY HOYNES, Chancellor's Professor of Economics and Public Policy, University of California, Berkeley
H. V. JAGADISH, Edgar F. Codd Distinguished University Professor of Electrical Engineering and Computer Science, University of Michigan
DANIEL KIFER, Professor of Computer Science, The Pennsylvania State University
SHARON LOHR, Professor Emerita, Arizona State University
NELA RICHARDSON, Senior Vice President and Chief Economist, ADP Research Institute
C. MATTHEW SNIPP, Burnet C. and Mildred Finley Wohlford Professor of Humanities and Sciences, Stanford University
ELIZABETH A. STUART, Hurley-Dorrier Professor and Chair of the Department of Biostatistics, Johns Hopkins Bloomberg School of Public Health

Study Staff

MELISSA CHIU, *Director*
JENNIFER PARK, *Study Director*
ALEX HENDERSON, *Senior Program Assistant*

COMMITTEE ON NATIONAL STATISTICS

KATHARINE G. ABRAHAM (*Chair*), Distinguished University Professor, University of Maryland, College Park
MICK P. COUPER, Research Professor, University of Michigan
WILLIAM A. DARITY, JR., Samuel DuBois Cook Professor of Public Policy, African and African American Studies, Economics, and Business, Duke University
ROBERT M. GOERGE, Senior Fellow, NORC of the University of Chicago
ERICA L. GROSHEN, Senior Economics Advisor, Cornell University
ROEE GUTMAN, Professor of Biostatistics, Brown University School of Public Health
COLLEEN M. HEFLIN, Professor of Public Administration and International Affairs, Syracuse University
DANIEL E. HO, William Benjamin Scott and Luna M. Scott Professor of Law, Professor of Political Science and Computer Science (by courtesy), Stanford University
HILARY HOYNES, Chancellor's Professor of Economics and Public Policy, University of California, Berkeley
H. V. JAGADISH, Edgar F. Codd Distinguished University Professor of Electrical Engineering and Computer Science, University of Michigan
SHARON LOHR, Professor Emerita, Arizona State University
LLOYD B. POTTER, State Demographer of Texas and Director of the Texas Demographic Center at University of Texas at San Antonio
NELA RICHARDSON, Senior Vice President and Chief Economist, ADP Research Institute
ELIZABETH A. STUART, Hurley-Dorrier Professor and Chair of the Department of Biostatistics, Johns Hopkins Bloomberg School of Public Health
FLORENCIA TORCHE, Edwards S. Sanford Professor of International Affairs and Sociology, Princeton University
SALIL VADHAN, Vicky Joseph Professor of Computer Science and Applied Mathematics, Harvard School of Engineering and Applied Sciences

Staff

MELISSA CHIU, *Director*
BRIAN HARRIS-KOJETIN, *Senior Scholar*
CONSTANCE F. CITRO, *Senior Scholar*

Reviewers

This Consensus Study Report was reviewed in draft form by individuals chosen for their diverse perspectives and technical expertise. The purpose of this independent review is to provide candid and critical comments that will assist the National Academies of Sciences, Engineering, and Medicine in making each published report as sound as possible and to ensure that it meets the institutional standards for quality, objectivity, evidence, and responsiveness to the study charge. The review comments and draft manuscript remain confidential to protect the integrity of the deliberative process.

We thank the following individuals for their review of this report:

MARY E. BOHMAN, Economic Research Service, U.S. Department of Agriculture (*retired*)
JOHN L. CZAJKA, Independent consultant, Bethesda, MD
NICK HART, Data Foundation
MICHAEL HORRIGAN, W.E. Upjohn Institute for Employment Research
FRAUKE KREUTER, Social Data Science Center, University of Maryland
JI-HYUN LEE, Department of Biostatistics and Cancer Quantitative Services, University of Florida
JERI MULROW, Data Solutions, Westat
AMY O'HARA, Federal Statistical Research Data Center, Georgetown University
CHARLES J. ROTHWELL, National Center for Health Statistics (*retired*)

Although the reviewers listed above provided many constructive comments and suggestions, they were not asked to endorse the conclusions or recommendations of this report, nor did they see the final draft before its release. The review of this report was overseen by **ALICIA L. CARRIQUIRY**, Department of Statistics, Iowa State University. This coordinator was responsible for making certain that an independent examination of this report was carried out in accordance with the standards of the National Academies and that all review comments were carefully considered. Responsibility for the final content rests entirely with the authoring committee and the National Academies.

Dedication

This Eighth Edition of *Principles and Practices for a Federal Statistical Agency (P&P)* is dedicated to the late Katherine K. Wallman (1943–2024). It is hard to overstate her contributions to the federal statistical system during her long career of federal service. She was a steadfast supporter of the Committee on National Statistics (CNSTAT), serving as a volunteer expert on study panels, regularly briefing the members and the statistical community at CNSTAT seminars, and citing *P&P* in statistical policy directives.

She began her career at the National Center for Education Statistics and then served in the chief statistician's office in the U.S. Office of Management and Budget (OMB) and in the Commerce Department. She left federal service in 1981 to become the first executive director of the newly formed Council of Professional Associations on Federal Statistics, established to speak for the value of federal statistics during an era of neglect. In 1992 she returned to federal service as the chief statistician and head of the OMB Statistical and Science Policy Office, a post she held until her retirement at the beginning of 2017.

Among her many accomplishments as chief statistician were overseeing the revision of Statistical Policy Directive No. 15 on race and ethnicity classification in 1997; the enactment of the Confidential Information Protection and Statistical Efficiency Act of 2002; the development of guidance from an Interagency Technical Working Group (which she co-chaired) that led to publication of the Supplemental Poverty Measure; and the issuance of Statistical Policy Directive No. 1 on Fundamental Responsibilities of Federal Statistical Agencies and Recognized Statistical Units in 2014.

Each of these initiatives and many others required her boundless energy, her steadfast dedication, and her legendary skills in bringing people and agencies with different viewpoints together for the common good. She remained active professionally after her retirement, working with the American Statistical Association and other organizations in support of the federal statistical system. It is with deep gratitude for her exceptional contributions to objective, relevant, and high-quality federal statistics for the public good that CNSTAT dedicates this eighth edition of *P&P* in her memory.

Acknowledgments

We thank the many people who contributed their time and expertise to the preparation of this report, including all the current members of CNSTAT. We are most appreciative of their cooperation and assistance.

We are particularly grateful to the CNSTAT staff, including board director Melissa Chiu, study director Jennifer Park, and senior program assistant Alex Henderson. Kirsten Sampson-Snyder and Bea Porter organized the review process, and Marc DeFrancis's thorough editing improved the readability and accessibility of the report. We are grateful to all of them for their contributions and help.

Finally, we thank the following federal agencies, which support the Committee on National Statistics directly and through a grant from the National Science Foundation, a cooperative agreement from the National Agricultural Statistics Service, and several individual contracts:

- National Science Foundation: Methodology, Measurement, and Statistics Program; National Center for Science and Engineering Statistics
- Social Security Administration: Office of Research, Evaluation, and Statistics
- U.S. Department of Agriculture: Economic Research Service, National Agricultural Statistics Service
- U.S. Department of Commerce: Bureau of Economic Analysis, U.S. Census Bureau
- U.S. Department of Education: National Center for Education Statistics

- U.S. Department of Energy: Energy Information Administration
- U.S. Department of Health and Human Services: Agency for Healthcare Research and Quality, National Center for Health Statistics, National Institute on Aging, Office of the Assistant Secretary for Planning and Evaluation
- U.S. Department of Housing and Urban Development: Office of Policy Development and Research
- U.S. Department of Justice: Bureau of Justice Statistics
- U.S. Department of Labor: Bureau of Labor Statistics
- U.S. Department of Transportation: Bureau of Transportation Statistics, and
- U.S. Department of the Treasury: Statistics of Income Division, Internal Revenue Service.

Without their support and their commitment to improving the national statistical system, the committee work that is the basis of this report would not have been possible.

Contents

Preface   xxi

Summary   1

1 Introduction   9
   AN ESSENTIAL AND COMPLEX SYSTEM, 10
   The Value of National Statistics, 10
   A Decentralized and Federated Statistical System, 10
   Unique Responsibilities, 13
   Expanded Availability of New Data Sources, 15
   THE PURPOSE OF THIS REPORT, 16
   OUR APPROACH, 16
   OVERVIEW, 17

2 The Value of National Statistics   19
   STATISTICS SUPPORT OUR NATIONAL INFRASTRUCTURE, 20
   STATISTICAL INFORMATION POWERS POLICYMAKING, 22
   Informs Political Representation, 22
   Informs Economic Decision Making, 22
   Informs Business Decisions, 23
   Assists Federal, State, and Local Government Action, 23
   Monitors the Social and Economic Health of the Nation, States,
      and Localities, 24
   Provides Evidence for Developing and Evaluating Public and
      Private-Sector Programs, 24

Provides Input to Social Science Research That Informs the
Public, 25
THE COSTS AND BENEFITS OF FEDERAL STATISTICS, 25

3 **Principles** 29
PRINCIPLE 1: RELEVANCE TO POLICY ISSUES AND
SOCIETY, 29
PRINCIPLE 2: CREDIBILITY AMONG DATA USERS AND
STAKEHOLDERS, 32
PRINCIPLE 3: TRUST AMONG THE PUBLIC AND DATA
SUBJECTS, 34
PRINCIPLE 4: INDEPENDENCE FROM POLITICAL AND
OTHER UNDUE EXTERNAL INFLUENCE, 37
PRINCIPLE 5: CONTINUAL IMPROVEMENT AND
INNOVATION, 41

4 **Practices** 45
PRACTICE 1: A CLEARLY DEFINED AND WELL-ACCEPTED
MISSION, 46
PRACTICE 2: NECESSARY AUTHORITY AND PROCEDURES
TO PROTECT INDEPENDENCE, 48
PRACTICE 3: COMMITMENT TO QUALITY AND
PROFESSIONAL STANDARDS OF PRACTICE, 52
PRACTICE 4: PROFESSIONAL ADVANCEMENT OF STAFF, 54
PRACTICE 5: AN ACTIVE RESEARCH PROGRAM, 57
Substantive Research and Analysis, 58
Research on Methodology and Operations, 58
Expanding the Statistical Use of Administrative Records, 60
Evaluating and Using Alternative Data Sources, 61
Value of an Active Research Program, 62
PRACTICE 6: STRONG INTERNAL AND EXTERNAL
EVALUATION PROCESSES FOR AN AGENCY'S
STATISTICAL PROGRAMS, 63
Evaluating Quality, Relevance, Efficiency, 63
Types of Reviews, 64
Sunsetting Statistical Products or Programs, 65
PRACTICE 7: COORDINATION AND COLLABORATION
WITH OTHER AGENCIES, 66
Coordinating Role of the Office of Management and Budget, 67
Forms of Interagency Collaboration, 67
International Collaborations, 69
Challenges and Rewards for Collaboration, 69

PRACTICE 8: RESPECT FOR DATA SUBJECTS AND DATA
   HOLDERS AND PROTECTION OF THEIR DATA, 71
Respecting Privacy in Surveys, 71
Protecting and Respecting the Autonomy of Human Research
   Participants, 72
Respecting the Holders and Subjects of Administrative and
   Other Data, 73
Protecting the Confidentiality of Data Subjects' Information, 73
PRACTICE 9: DISSEMINATION OF STATISTICAL PRODUCTS
   THAT MEET USERS' NEEDS, 75
Public Statistical Data Products, 77
Restricted-Access Statistical Products, 78
Recent Innovations in Facilitating Access to Confidential
   Data for Statistical Purposes, 79
PRACTICE 10: OPENNESS ABOUT SOURCES AND
   LIMITATIONS OF THE DATA PROVIDED, 81

Acronyms and Abbreviations   85

Glossary of Selected Terms   89

References   97

Appendix A Legislation, Regulation, and Guidance Governing
Federal Statistics (online only)   111

Appendix B Organization of the U.S. Federal Statistical System
(online only)   159

Appendix C Additional Frameworks Relevant for Federal Statistics
(online only)   191

Appendix D Biographical Sketches   207

Boxes, Figures, and Table

BOXES

S-1 Principles and Practices, 6

1-1 Principles and Practices, 18

2-1 Federal Statistics as a Public Good, 27

3-1 Summary: Principle 1 (Relevance), 32
3-2 Summary: Principle 2 (Credibility), 34
3-3 Summary: Principle 3 (Trust), 37
3-4 Summary: Principle 4 (Independence), 41
3-5 Summary: Principle 5 (Innovation), 44

B-1 Interagency Council on Statistical Policy Membership, 179

C-1 Interagency Council on Statistical Policy Principles for Using Non-Statistical Data for Statistical Purposes, 193
C-2 Federal Data Ethics Tenets, 194
C-3 American Statistical Association Ethical Guidelines for Statistical Practice, 195
C-4 American Economic Association Principles of Economic Measurement, 196
C-5 American Association for Public Opinion Research Code of Professional Ethics and Practices, 197

C-6 Office of Science and Technology Policy Blueprint for an AI Bill of Rights, 198
C-7 United Nations Fundamental Principles of Official Statistics, 199
C-8 International Statistical Institute Declaration on Professional Ethics, 201
C-9 European Statistics Code of Practice, 202
C-10 Organisation for Economic Co-operation and Development Good Statistical Practice, 203

FIGURES

1-1 The U.S. federal statistical system as depicted on StatsPolicy.gov, 11

B-1 The U.S. federal statistical system as depicted on StatsPolicy.gov, 161
B-2 Recognized statistical agencies and units by congressional appropriations committee and parent Chief Financial Officer Act agency, 188

C-1 Generic statistical business process model, 205

TABLE

A-1 Significant Legislation, Regulations, and Guidance Supporting Federal Statistics, by Key Principle(s), 112

Preface

This report is a flagship publication of the Committee on National Statistics (CNSTAT). CNSTAT is a standing unit of the National Academies of Sciences, Engineering, and Medicine, established in 1972 to provide an independent, objective resource for evaluating and improving the work of the decentralized federal statistical system. Over its 52-year history, under the terms of the 1863 congressional charter to the National Academy of Sciences to provide advice to the government on scientific and technical matters, CNSTAT has produced 340 reports evaluating federal statistical programs, surveys, and statistical methods for the public good, regardless of persuasion or party (Committee on National Statistics, 2024).

The first edition of *Principles and Practices for a Federal Statistical Agency* (*P&P*; also known as "the purple book") was published in 1992. During legislative debates regarding the (unsuccessful) establishment of a Bureau of Environmental Statistics and the (successful) establishment of a Bureau of Transportation Statistics, congressional staff asked CNSTAT for advice on what constitutes an effective federal statistical agency. CNSTAT prepared a document providing high-level guidance. That document, the first edition of *P&P*, defined and discussed reasons for the establishment of a statistical agency, identified three fundamental principles for an effective statistical agency (relevance to policy, credibility with data users, and trust of data subjects), and identified 11 practices to enable a statistical agency to put these principles into action and adhere to them. The report has proven helpful to Congress, the Government Accountability Office (GAO), the Office of Management and Budget, federal statistical agencies, and others seeking to understand what constitutes an effective and credible statistics entity.¹ *P&P* is designed to assist them, as well as the statistical agencies' leadership, their staff, and the research organizations that support them, to be fully aware of the standards and ideals that are fundamental to the agencies' work.

Since the first edition of *P&P*, several other national and international statistical organizations, including the American Statistical Association and the United Nations Statistical Commission, have issued guidance that aligns with, echoes, and reinforces many of the same themes (see Appendix C). Many of the principles and practices elaborated here also apply to statistical activities in federal data strategy, evaluation,² and program agencies; in state, local, and tribal government agencies; and in nongovernmental organizations. Indeed, the past few years have witnessed greater attention to using administrative and private-sector data sources, not only for national statistics (National Academies of Sciences, Engineering, and Medicine [NASEM], 2017a,b,c, 2023b,c, 2024c) but also more broadly for program evaluation and evidence-based policymaking (Commission on Evidence-Based Policymaking, 2017).

Now more than ever, at a time of change in the U.S. statistical system, this report has an important role. It provides an independent perspective that spans departments and administrations. To maintain its usefulness to policymakers, the report has been updated every 4 years to provide a current edition to newly appointed cabinet secretaries and other personnel at the beginning of each presidential administration or second term.³

CNSTAT has made some changes to *P&P* over time. The first three editions included three principles (National Research Council [NRC], 1992, 2001, 2005). In light of growing concerns about threats to the agencies' independence, the fourth edition elevated statistical agency independence from a practice to a fourth principle (NRC, 2009). The number of practices fluctuated across the first seven editions, as conclusions and recommendations in CNSTAT study reports led to adding or combining some practices. The seventh edition added a fifth principle on Continual Improvement and Innovation, which has been a strong theme in a number of practices, to recognize its importance for the effective functioning of statistical agencies in the 21st century (NASEM, 2017a, 2021a, 2023b,c).

The principles and practices described in this eighth edition are largely unchanged from the prior editions. Language originating from prior editions is not marked as such for ease of reading. However, examples have been updated to reflect legal, policy, and programmatic changes since the

---

¹See (Citro, 2014; GAO, 1995, 2007, 2012; OMB, 2007, 2014b).
²See (OMB 2020a) in Appendix A and *Principles and Practices for Federal Program Evaluation* (NASEM, 2017d).
³Beginning with the second edition in 2001.

prior volume. In addition, the introductory chapter from the seventh edition has been split into two chapters and highlight boxes have been added to chapters to improve the clarity of key messages. The appendices (available only online) have been substantially updated to reflect the legal, regulatory, and policy changes to the federal statistical system that have resulted from the Evidence Act of 2018, as well as the expansion of guidelines for statistical practice that have emerged both nationally and internationally.

Katharine G. Abraham, *Chair*
Committee on National Statistics
October 2024

Summary

Highlights

- Public trust in federal statistics and in the agencies that provide them is essential to a shared understanding of our economy and society.
- Articulating agreed-upon principles and practices provides a common foundation to producers and users of federal statistics.
- In 1992, the board members of the Committee on National Statistics were first tasked to issue guidance for the federal statistical system as whole.
- Five principles guide federal statistical agency programs.
- Ten practices support achievement of these principles.
- These principles and practices are largely unchanged from prior editions.
- Modest revisions in this eighth edition reflect evolution in data ecosystems and significant federal statistical policy changes observed or anticipated since the passage of the Evidence Act of 2018.

THE VALUE OF NATIONAL STATISTICS

Statistics are essential not only for policymakers and program administrators at all governmental levels, but also for individuals, households, businesses, and other organizations to make informed decisions and for scientists to add to knowledge. Even more broadly, the effective operation of a democratic system of government depends on the unhindered provision of impartial, scientifically based statistical information to its public on a wide range of issues, including employment, growth in the economy, the cost of living, crime victimization, family structure, physical and mental health, educational attainment, energy use, and the environment. Furthermore, this flow of trustworthy information is critical to shaping citizens' understanding when confronted with today's avalanche of statistics in traditional and social media.

A DECENTRALIZED SYSTEM

In the United States, there are more than a dozen federal statistical agencies[1] whose principal function is to collect, compile, analyze, and disseminate information for such statistical uses as monitoring key economic and societal indicators, informing policy, allocating legislative seats and government funds, evaluating programs, and conducting scientific research. Federal statistical agencies often rely on data provided by other federal agencies and by state, local, and tribal governments to produce official statistics.

Statistical agencies have unique responsibilities. Although statistical agencies provide objective and impartial information that informs policymakers, they should not advocate policies or take partisan positions that would undercut public trust and the credibility of the statistics they produce. In addition, recent statutory and regulatory changes have expanded the responsibilities of federal statistical agencies and other federal agencies to include advising departments on the use of statistical data for evidence-based policymaking.

A COMMON FOUNDATION

Communicating the unique responsibilities of federal statistical agencies—and best practices on achieving them—remains essential to a

---

[1] The United States has a decentralized statistical system. There are thirteen recognized statistical agencies and three recognized statistical units. In addition, using a threshold definition as $3 million in estimated or direct funding for statistical activities in the forthcoming or either of the past two fiscal years, there were more than 109 other agencies that conducted statistical activities in 2022 (Office of Management and Budget [OMB], 2023b, 2024c).

well-functioning statistical system. To this end, the Committee on National Statistics (CNSTAT) of the National Academies of Sciences, Engineering, and Medicine articulates five principles and 10 practices as part of its mission to provide an independent review of federal statistical activities.

From its first edition in 1992 to this eighth edition, *Principles and Practices for a Federal Statistical Agency (P&P)*[2] is intended to support the invaluable role of relevant, credible, trusted, independent, and innovative government statistics. With modest refinement over the years, the principles and practices have endured. They have influenced federal statistical policy in substantial and lasting ways. Indeed, acts of Congress as well as regulations and statistical policy directives issued by the OMB have codified many of these principles and practices. The volume is used across federal statistical agencies and an ever-expanding community of producers and users of national statistics to introduce and support their staff in a shared culture.

Five Principles

Federal statistical agencies are coordinated by OMB and are subject to government regulations and guidance, but their missions and contributions to the public good are best seen as resting on five well-established and fundamental principles. In brief, these principles are relevance, credibility, trust, independence, and innovation. These are coequal, not ranked, and are intended to reinforce one another in practice. They are summarized below, listed in Box S-1, and described in greater detail in Chapter 3.

Principle 1: Relevance to Policy Issues and Society

Federal statistical agencies must provide objective, accurate, and timely information that is relevant to important public policy issues. They must also help ensure that these products are on the table when decisions are made. To develop relevant statistics needed by policymakers in Congress, the executive branch, and other entities, statistical agencies must have a solid understanding of the public policy issues, federal programs, and information needs in their domains. To ensure that they are providing relevant information, statistical agencies need to reach out to a wide range of their data users, including staff in their own departments and other federal departments who use their data, members of Congress and their staffs, state and local government agencies, academic researchers, businesses, and members of the general public, including local organizations. To facilitate best use of their products, statistical agencies must educate users and adapt dissemination to their users' needs. However, statistical agencies should be

---

[2]We refer to this report as *P&P* throughout the remainder of the text.

careful not to become involved with policy development or implementation, as those activities could affect their ability (or the perception of their ability) to conduct impartial and objective statistical activities.

Principle 2: Credibility Among Data Users and Stakeholders

Federal statistical agencies must have credibility with those who use their data and information. The value of statistical agencies rests fundamentally on the accuracy of their data products. Because few data users have the resources to verify the accuracy of statistical information, users rely on an agency's reputation for disseminating high-quality, objective, and useful statistics in an impartial manner. Agencies build and maintain credibility through clear public commitments to professional practice and transparency in all that they do, including informing users of the strengths and weaknesses of their data. Agencies should communicate the "fitness for use" of their statistical products as meeting the degree of accuracy required for a particular, responsible use.

As alternative data sources are used to supplement (or sometimes replace) survey data, federal statistical agencies must continually strive for verifiable improvements in accuracy and timeliness in the estimates they produce. This can include reduction of known bias (such as from the use of administrative data blended with survey data) measured by the reduction in measures of total error (National Academies of Sciences, Engineering, and Medicine [NASEM], 2023c).

Principle 3: Trust Among the Public and Data Subjects

Federal statistical agencies must have the trust of those whose information they obtain. Because virtually every person, household, business, state or local government, and organization is the subject of some federal statistics, public trust is essential for the continued effectiveness of federal statistical agencies. Individuals and entities providing data directly or indirectly to federal statistical agencies must trust that the agency is collecting information that serves a public purpose, and that the agency will appropriately handle and protect their information. Federal statistical agencies not only have legal and ethical obligations that require them to fulfill these expectations, but they also have the obligation to effectively communicate the value of the data they collect and the methods they use for obtaining and protecting them. An effective statistical agency has policies and practices to instill the highest possible commitment to professional ethics among its staff and builds a culture of protecting the confidentiality of its data and engendering respect for those who provide data.

Principle 4: Independence from Political and Other Undue External Influence

Federal statistical agencies must be independent from political and other undue external influence in developing, producing, and disseminating statistics. Statistical agencies must be impartial and execute their missions without being subject to pressures to advance any political or personal agenda. They must avoid even the appearance that their collection, analysis, and reporting processes might be manipulated for political or other purposes or that individually identifiable data might be obtainable for nonstatistical purposes. Only in this way can statistical agencies serve as trustworthy sources of objective, relevant, accurate, and timely information. Protection from undue outside influences requires that statistical agencies have the authority to make professional decisions concerning their programs, including authority over the selection and promotion of staff; the processing, secure storage, and maintenance of data; and the timing and content of data releases, accompanying press releases, and documentation.

Principle 5: Continual Improvement and Innovation

Federal statistical agencies must continually seek to improve and innovate their processes, methods, and statistical products to better measure an ever-changing world. Federal statistical agencies and programs cannot be static but must continually work to create reliable information on new policy questions, adopt improvements in all aspects of their operations, and respond to user demands for more timely and granular information. An effective statistical agency not only seeks out and evaluates potential new data sources that could provide useful information, but also tests and implements new methods to enhance the accuracy and cost-effectiveness of its data collection, processing, and dissemination processes. It also seeks ways to reduce burden on data subjects. It works closely with data users to identify potential new and useful statistical products.

Ten Practices

To fulfill these five principles, 10 practices are essential for statistical agencies to adopt. These practices represent the ways and means of making the basic principles operational and facilitating an agency's adherence to them. Practices 1 to 4 pertain to an agency's operations, internally and within the federal government; practices 5 to 7 bridge internal operations and external relations with the professional statistical and research communities; and practices 8 to 10 focus externally on an agency's key

> **BOX S-1**
> **Principles and Practices**
>
> Five principles guide federal statistical agency programs:
>
> 1. Relevance to Policy Issues and Society
> 2. Credibility Among Data Users and Stakeholders
> 3. Trust Among the Public and Data Subjects
> 4. Independence from Political and Other Undue External Influence
> 5. Continual Improvement and Innovation
>
> Ten practices support achievement of these principles:
>
> 1. A Clearly Defined and Well-Accepted Mission
> 2. Necessary Authority and Procedures to Protect Independence
> 3. Commitment to Quality and Professional Standards of Practice
> 4. Professional Advancement of Staff
> 5. An Active Research Program
> 6. Strong Internal and External Evaluation Processes for an Agency's Statistical Programs
> 7. Coordination and Collaboration with Other Agencies
> 8. Respect for Data Subjects and Data Holders and Protection of Their Data
> 9. Dissemination of Statistical Products That Meet Users' Needs
> 10. Openness About Sources and Limitations of the Data Provided

constituents: data users, data subjects, and data holders.[3] The 10 practices are listed in Box S-1 and detailed in Chapter 4.

THE CONTRIBUTION OF THIS REPORT

The preparation of this report is unique among other National Academies reports. Because *P&P* is intended to reflect and uphold the most salient guidance for the federal statistical system as a whole, the authoring committee is CNSTAT itself, due to its role as an independent, objective, and centralizing influence on the system. CNSTAT members include experts in statistical and computational methods, survey research, economic, social and demographic measurement, administrative law, and other relevant fields. Additionally, unlike many other National Academies reports, this report does not include recommendations. Instead, this report provides

---

[3] The terms *data subject* and *data holder* in this edition have replaced the term *data provider* used in the prior edition to reflect greater precision when describing entities contributing or facilitating access to data. See glossary.

principles and practices, which serve as guidelines and not prescriptions. By adhering to the principles and following the practices, a federal statistical agency will be well positioned to provide the relevant, credible, trusted, independent, and innovative statistical information that the public requires.

Since 2001, CNSTAT has issued a new edition of P&P to coincide with the start of each presidential term. This serves two key functions: to communicate with a new audience these principles and practices, and to account for the expected, ongoing changes in society and the federal statistical system. The principles and practices described in this edition remain unchanged since the prior edition. The updates made to this volume instead consider how changes in the federal statistical system intersect with these longstanding guidelines. Throughout, the volume accounts for the wide-ranging observed and anticipated impacts of the Foundations for Evidence-Based Policymaking Act of 2018 (2019), which expanded the role of heads of the recognized statistical agencies and units in their departments; expanded the responsibilities of the Chief Statistician in coordinating expanded data sharing across agencies and access to confidential data for evidence building; and prescribed an enlarged role for federal surveys and administrative records to be used in support of sound policymaking. This edition has added a glossary and tightened terminology to improve accessibility to an ever-widening audience, and its appendices have been extensively updated and rearranged for ease of reference.

CNSTAT intends for these principles and practices to assist statistical agencies and units, the research organizations supporting them, as well as other government agencies nationwide engaged in statistical activities, to carry out their responsibilities to provide accurate, timely, relevant, and objective information for public and policy use. It also intends this report to inform decision makers, data users, and others about the characteristics of statistical agencies that enable them to serve the public good.

1

Introduction

Highlights

- The nation relies on high-quality statistical information.
- The nation has a decentralized statistical system. Within the federal government, there are 16 recognized statistical agencies and units, and over 100 statistical programs. States, localities, and tribal governments provide essential data for federal statistics. They and other groups are important users of federal data.
- Recent changes in federal law expanded the responsibilities of federal statistical agencies and established new roles for Statistical Officials in other federal agencies.
- Technical methods to develop and link new data sources have expanded across disciplines and applications.
- A common foundation for defining expectations for federal statistical practice is necessary.
- With periodic updates, this report has long helped to unify the federal statistical system. It remains an important tool to communicate shared expectations and to guide statistical policymaking.
- The board members of the Committee on National Statistics were tasked by the National Academies of Sciences, Engineering, and Medicine to review the prior edition of this report and

> determine if revisions were necessary to reflect current federal policy and national interests.
> - The board determined that the principles and practices of the prior edition should remain intact, although examples and references should be updated and the overall text reorganized to improve accessibility. This eighth edition reflects these modest changes.

AN ESSENTIAL AND COMPLEX SYSTEM

A number of features of the U.S. statistical system have remained constant across decades—the high value of national statistics to our society, the decentralized and federated nature of statistical production, and the unique responsibilities entrusted to federal statistical agencies. Yet, our national data infrastructure also has changed over time to reflect the needs of society. Some recent changes have affected the organization and expansion of the system, access to federal statistics, and the potential impact of federal statistical agencies on evidence-making.

The Value of National Statistics

National statistics help the public shape their country. They help individuals and households to make decisions about where to live, work, and attend school. National statistics inform decisions of businesses and other organizations about market changes and opportunities. They provide essential information for policymakers and program administrators at all governmental levels to identify the public's needs and interests as well as the efficacy of public policies. National statistics are a resource to researchers and scientists. Even more broadly, the effective operation of a democratic system of government depends on the unhindered flow of impartial, scientifically based statistical information to its citizens on a wide range of issues. These issues range from employment, growth in the economy, and the cost of living to crime victimization, family structure, physical and mental health, educational attainment, and energy use and the environment.

A Decentralized and Federated Statistical System

In the United States, the national statistical system is highly decentralized. Rather than being managed by a single government agency (as is the case in many other countries), the responsibility for producing federal statistics is shared across several specialized federal agencies. Further, many of these federal agencies rely on data from state, local, and tribal governments as well as other entities to produce national statistics.

This section summarizes the key features of the U.S. national statistical system. See Figure 1-1. Authorities, agency membership, structure, and work processes are discussed in detail in Appendices A and B, available in the online copy of this report.

The Office of the Chief Statistician of the United States

Through the director of the Office of Management and Budget (OMB), the Chief Statistician of the United States has the responsibility of coordinating domestic and international federal statistical policy (Foundations for Evidence-Based Policymaking Act of 2018, 2019; Paperwork Reduction Act, 1995).

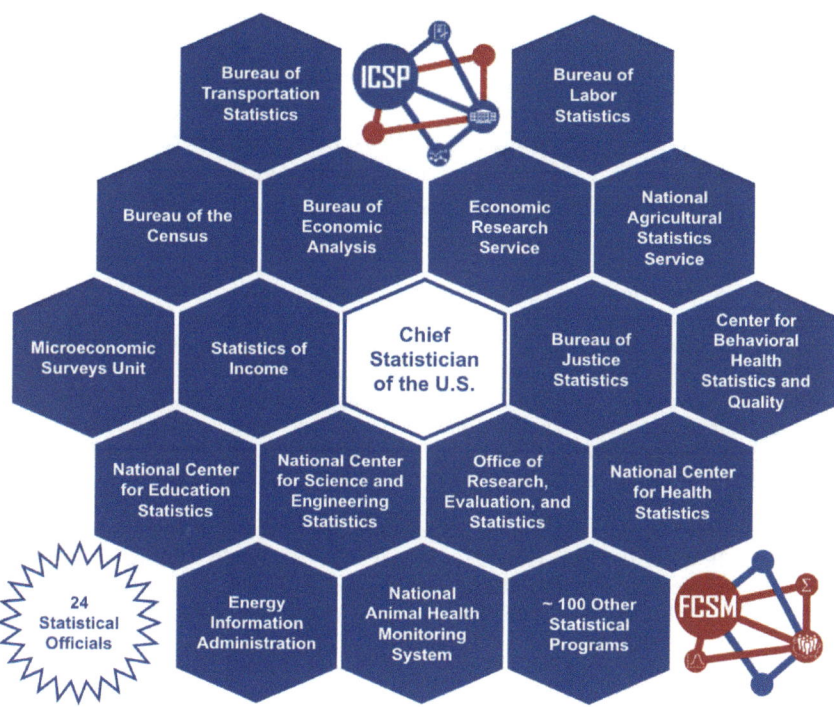

FIGURE 1-1 The U.S. federal statistical system as depicted on StatsPolicy.gov.
SOURCE: https://www.statspolicy.gov/about/#office-chief

Federal Statistical Agencies and Units

In the United States, the responsibility for collecting the necessary data to generate statistics for public use is distributed among a set of agencies and organizational units in the executive branch. A federal statistical agency or unit is defined in law as "[...] An agency or organizational unit of the executive branch whose activities are predominantly the collection, compilation, processing, or analysis of information for statistical purposes" (Foundations for Evidence-Based Policymaking Act of 2018, 2019).

At present, there are 13 recognized statistical agencies[1] and three recognized statistical units[2] located in cabinet departments and independent agencies.[3] In addition, over 100 program, policy, and research agencies have sizable statistical activities, many of which feed into the production of statistics from the recognized agencies (OMB, 2023b).

State, Local, and Tribal Agencies

State, local, and tribal agencies collect data as part of the administration of federal programs such as Temporary Assistance to Needy Families, Unemployment Insurance, the Supplemental Nutrition Assistance Program (SNAP or foodstamps), Medicaid, highway funding, and energy assistance, among others. In addition, there are extensive federal-state cooperative networks that enable the production of essential federal statistics, such as births and deaths;[4] K–12 education;[5] employment and wages;[6] and health.[7]

---

[1] The U.S. recognized federal statistical agencies are (13): Bureau of Economic Analysis (Department of Commerce); Bureau of Justice Statistics (Department of Justice); Bureau of Labor Statistics (Department of Labor); Bureau of Transportation Statistics (Department of Transportation); Census Bureau (Department of Commerce); Economic Research Service (Department of Agriculture); Energy Information Agency (Department of Energy); National Agricultural Statistics Service (Department of Agriculture); National Center for Education Statistics (Department of Education); National Center for Health Statistics (Department of Health and Human Services); National Center for Science and Engineering Statistics (National Science Foundation); Office of Research, Evaluation, and Statistics (Social Security Administration); and Statistics of Income (Department of the Treasury; Foundations for Evidence-Based Policymaking Act of 2018, 2019).

[2] There are three recognized federal statistical units: Microeconomic Surveys Unit (Federal Reserve Board); Center for Behavioral Health Statistics and Quality (Substance Abuse and Mental Health Services Administration, Department of Health and Human Services); and National Animal Health Monitoring System (Animal and Plant Health Inspection Service, Department of Agriculture; Foundations for Evidence-Based Policymaking Act of 2018, 2019).

[3] An organizational chart is included in Appendix B.
[4] https://www.cdc.gov/nchs/nvss/index.htm
[5] https://nces.ed.gov/ccd/
[6] https://www.bls.gov/cew/; https://www.bls.gov/oes/; https://lehd.ces.census.gov/
[7] https://www.bls.gov/iif/overview/soii-overview.htm; https://www.cdc.gov/mmwr/index.html

These data are compiled by federal agencies and used to describe the operation of programs, evaluate programs, inform policy and the public, and support budgeting. They also feed into important federal-state statistical programs, such as vital statistics on births and deaths.

Private Organizations

Federal statistical agencies cannot collect all of the data that they seek on their own. Many private organizations conduct data collection and analysis activities of interest to federal statistical agencies. The organizations can be for-profit, nonprofit, and/or academically affiliated. For example, there are research organizations that assist federal data collection and production. Additionally, there are firms that collect data for commercial purposes, such as credit bureaus, insurance companies, banks, and payroll companies. There are also commercial data brokers that collect, buy, and aggregate data, which they can sell to researchers or statistical agencies. Data provided by such private sources are increasingly important inputs to the federal statistical system (see Expanded Availability of New Data Sources, below). In some cases, these data products fill (whether in perception or reality) data user needs that are unmet by federal statistical agencies. To responsibly make use of these data products, federal statistical agencies must carefully examine the integrity of the data provided by such private organizations.

Unique Responsibilities

Federal statistical agencies and units have unique responsibilities that set them apart from other federal agencies. Although the subject matter varies, each statistical agency has a mission to acquire, produce, and disseminate information for statistical purposes. Because the value of national statistics is great, the definition and, subsequently, the responsibilities of federal statistical agencies are established in law. As the demand for national statistics in supporting evidence-based policymaking grows, new responsibilities and roles have been added in law, and more are expected to be developed in future regulation.

Historical Responsibilities

The Confidential Information Protection and Statistical Efficiency Act of 2002 defined a federal statistical agency or unit as "[…] an agency or organizational unit of the executive branch whose activities are predominantly the collection, compilation, processing, or analysis of information for statistical purposes." (See Confidential Information Protection and Statistical Efficiency Act, 2002, Sec 2 (8).[8])

The Foundations for Evidence-Based Policymaking Act of 2018, also known as the Evidence Act, codified the following set of responsibilities[9]:

1. Produce and disseminate relevant and timely statistical information;
2. Conduct credible and accurate statistical activities;
3. Conduct objective statistical activities; and
4. Protect the trust of information providers by ensuring the confidentiality and exclusive statistical use of their responses.[10]

Importantly, "statistical purpose or use" was also defined in law. It (a) means the description, estimation, or analysis of the characteristics of groups, without identifying the individuals or organizations that comprise such groups; and (b) includes the development, implementation, or maintenance of methods, technical or administrative procedures, or information resources that support the purposes described in subparagraph A (Confidential Information Protection and Statistical Efficiency Act, 2002). In contrast, a "nonstatistical purpose" is defined as using data in identifiable form for such purposes as "administrative, regulatory, law enforcement, adjudicatory, or other purpose that affects the rights, privileges, and benefits of a particular, identifiable respondent."[11]

As an example, consider information that might be collected on a person's income by a federal agency. A statistical agency or unit would only collect and use that information to compute statistics such as median income, the percentage of families below the poverty line, or the percentage eligible for SNAP benefits. A program agency might collect and use that information to determine whether that individual or family was eligible to receive SNAP or other benefits, and then grant or deny those benefits based upon that information. This latter use would be a nonstatistical purpose, and statistical agencies are generally prohibited by law from using their data in this manner.

Expanded Responsibilities of Statistical Officials

As discussed throughout this report, the Evidence Act made many changes in roles and responsibilities for evidence-building and the generation

---

[8]Appendix A provides an extensive overview of the statutory, regulatory, and policy basis guiding U.S. federal statistical practice, which dates back to the 1930s.

[9]These responsibilities codified OMB Statistical Policy Directive No. 1 as part of the Evidence Act (OMB, 2014b).

[10]44 USC § 3563(a)(1).

[11]44 USC § 3561(8)(a).

of statistics in the federal government. Among the most impactful for the statistical agencies may be their expanded role within their respective departments. This has given them and their heads (as Statistical Officials for their departments) new responsibilities and opportunities related to the statistical use of data for providing evidence for program evaluation, working in many departments and agencies with the people named as Chief Data Officers and Evaluation Officers under the Act (OMB, 2024e). So, too, the Evidence Act expanded the role of the Chief Statistician in coordinating expanded data sharing among agencies and access to confidential data for evidence building.

Because the Evidence Act requires implementing regulations and guidance from OMB, only some of which have been issued as of the writing of this volume (see Appendix A), many details regarding the roles of the new officers and their relationships with the statistical agency heads are yet to be determined. To achieve the long-term goals of greater use of government administrative and survey data for statistical purposes and for the public good, the broader statistical, research, and evaluation communities will need to come together in a productive dialogue. This dialogue will determine the appropriate principles and practices for these expanding statistical and evaluation activities and for the programs and units within which they are occurring, with potential impact on future editions of this report.

Expanded Availability of New Data Sources

Over the past decade, declining survey response rates have contributed to an increased need to use administrative data, private-sector data, and other sources of uncurated data to supplement or, in some cases, replace existing survey systems to produce federal statistics. Part D of the Evidence Act envisions that, unless expressly prohibited by law, federal statistical agencies generally should have access to federal administrative data for statistical purposes. More recently, interest has grown in linked or blended data—that is, combining data files initially collected to be used separately. The advantages and disadvantages of using data initially collected for nonstatistical purposes to produce statistics, as well as linked or blended data, have been well documented (National Academies of Sciences, Engineering, and Medicine [NASEM], 2017a,b,d, 2023b,c, 2024c).[12] Benefits include efficiency, robustness of data files, and reduction of public burden. Challenges include the assessment of quality, equity, and interoperability of data and measures, reliability of data sources (including potential cost),

---

[12]See, for example, https://www.statspolicy.gov/assets/docs/ICSP-The%20Use%20of%20Private%20Datasets%20by%20Federal%20Statistical%20Programs-1-6-2023.pdf

transparency, confidentiality,[13] and (potentially) concerns over informed consent. Ultimately, navigating responsible statistical use of new data sources within their unique statutory requirements poses further challenges to federal statistical agencies within a changing national data infrastructure.

THE PURPOSE OF THIS REPORT

This report is a tool to communicate the unique responsibilities of federal statistical agencies and to provide guidance to their staffs on how to achieve them. It does this by providing a common foundation in the form of principles and practices, referencing each to responsibilities established by law, regulation, and/or policy. The report distinguishes between "principles," which are fundamental and intrinsic to the concept of a federal statistical agency, and "practices," which are ways and means of making the basic principles operational and facilitating an agency's adherence to them.

The principles and practices in this report remain guidelines, not prescriptions. They are aspirational in nature to foster better decision making. The report is intended to assist statistical agencies and units, as well as other agencies, including state and local agencies, as they engage in statistical activities, and to inform legislative and executive branch decision makers, data users, and others about the characteristics of statistical agencies that enable them to serve the common good.

OUR APPROACH

The preparation of this report is unique among National Academies reports. Because *P&P* is intended to reflect and uphold the most salient guidance for the federal statistical system as a whole, the authoring committee is the Committee on National Statistics (CNSTAT) itself, due to its role serving as an independent, objective, and centralizing influence on the system.

Additionally, the periodicity of this report is unique. Since issuing the second edition in 2001, CNSTAT has issued an edition to coincide with the start of each new presidential term. This serves two key functions: to communicate with a new audience regarding these principles and practices,

---

[13]Note that the Confidential Information and Statistical Efficiency Act of 2002, Title V of the E-Government Act and reauthorized in 2018 as Title III of the Evidence Act, requires that information acquired (including administrative or other alternative data sources) under a pledge of confidentiality and for exclusively statistical purposes shall only be used for statistical purposes. This information shall not be disclosed in identifiable form without the informed consent of the respondent (Confidential Information and Statistical Efficiency Act of 2018, 2019).

and to account for the expected, ongoing changes in society and the federal statistical system.

Accordingly, for this eighth edition, CNSTAT was charged with the following statement of task:

> The National Academies of Sciences, Engineering, and Medicine will convene an ad hoc committee to update *Principles and Practices for a Federal Statistical Agency*, which provides advice on what constitutes an effective federal statistical agency. The report will take into account changes in laws, regulations, and other aspects of the environment for federal statistical agencies that have taken place since the release of the seventh edition.

To conduct its work, CNSTAT reviewed prior editions of *P&P* and took note of recent changes in the statutory, regulatory, and policy environment as well as the data user community. The form of *P&P* is unique in another manner. Unlike many other National Academies reports, the recommendations of this report are articulated in the principles and practices as guidelines, not prescriptions.

The principles and practices described in this edition remain largely unchanged from the prior edition. The updates made to this volume instead consider how changes in the federal statistical system and the environment in which it operates intersect with these long-standing guidelines. Throughout, the volume accounts for the wide-ranging observed and anticipated impacts of the Foundations for Evidence-Based Policymaking Act of 2018, which expanded the role of heads of the recognized statistical agencies and units in their departments and prescribed an enlarged role for federal surveys and administrative records to be used in support of sound policymaking. A glossary and tightened terminology have been added to improve accessibility to an ever-widening audience. The appendices have been extensively updated and rearranged for ease of reference.

OVERVIEW

After this Introduction, Chapter 2 provides an overview of the value of national statistics. Chapter 3 presents five basic interrelated principles that statistical agencies must embody to carry out their mission fully. Chapter 4 discusses 10 important practices that provide the means for statistical agencies to implement the five principles. The first four practices pertain to an agency's operations, internally and within the federal government, while practices 5 through 7 bridge internal operations and external relations with the professional statistical and research communities, and practices 8 through 10 face externally, toward an agency's key constituents: data users,

> **BOX 1-1**
> **Principles and Practices**
>
> Five principles guide federal statistical agency programs:
>
> 1. Relevance to Policy Issues and Society
> 2. Credibility Among Data Users and Stakeholders
> 3. Trust Among the Public and Data Subjects
> 4. Independence from Political and Other Undue External Influence
> 5. Continual Improvement and Innovation
>
> Ten practices support achievement of these principles:
>
> 1. A Clearly Defined and Well-Accepted Mission
> 2. Necessary Authority and Procedures to Protect Independence
> 3. Commitment to Quality and Professional Standards of Practice
> 4. Professional Advancement of Staff
> 5. An Active Research Program
> 6. Strong Internal and External Evaluation Processes for an Agency's Statistical Programs
> 7. Coordination and Collaboration with Other Agencies
> 8. Respect for Data Subjects and Data Holders and Protection of Their Data
> 9. Dissemination of Statistical Products That Meet Users' Needs
> 10. Openness About Sources and Limitations of the Data Provided

data subjects, and data holders. Chapters 3 and 4 include commentary on each principle and practice.

Three appendices are provided as a resource in the online edition. Appendix A summarizes the history of federal legislation, regulations, and executive policy memoranda and other guidance corresponding to the principles contained in this report. Appendix B summarizes the history and organization of the federal statistical system, recognized statistical agencies and units, Statistical Officials in other federal agencies, and the coordinating function of OMB. Appendix C summarizes several national and international frameworks for guiding official statistics.

2

The Value of National Statistics

Highlights

- Similar to the physical infrastructure of a nation, such as its interstate highways, national defense assets, and interstate utility grids, a national data infrastructure connects and serves society.
- The statistical information that makes up our national data infrastructure powers decision making. It informs political representation and business decisions. Statistical information allows policymakers to monitor the economic and social health of the nation, states, and localities. It provides evidence for developing and evaluating public- and private-sector programs. It informs federal, state, and local government action. Statistical information also provides input to research that contributes to public understanding of our world.
- Federal statistics are a sound national investment. The cost of federal statistical programs is a tiny fraction of overall U.S. federal spending. The annual benefit to our nation has been valued at $778.0 billion in 2022.[a]

---

[a] See https://www.commerce.gov/news/blog/2024/12/revenue-industries-heavily-reliant-us-government-data. This paragraph and footnote were modified after release of the report to provide a more precise estimate of value and its context based on publication released on December 3, 2024.

STATISTICS SUPPORT OUR NATIONAL INFRASTRUCTURE

People in the United States and other countries rely on data and statistics to live their lives, often without realizing it. They may check weather, traffic, or air quality reports and other readily available data to guide how they go about their day. They may use data to inform key family and personal decisions, such as where to live, based on information about housing, crime, schools, and jobs. In their own jobs, people may use data to guide policies and programs, make investment decisions, plan for the future, and develop knowledge.

People in a democracy also rely on accurate and trustworthy information to carry out their civic duties and maintain a government that protects and serves their interests. An informed citizenry must judge the merits of government actions through periodic votes for elected officials, and to become informed citizens depend on widely distributed information available both directly and through the media. Lawmakers require accurate population statistics for drawing up legislative districts that give all voters an equal voice. Fiscal and monetary policy actions rely heavily on economic releases. Identifying problems to address and opportunities to pursue requires that decision makers in both the governmental and private sectors have objective and timely information on the society and economy. Adding to knowledge about the society and the economy, in turn, requires detailed information for researchers to analyze in a wide range of fields.

The cornucopia of statistical information that people use in all these ways and often take for granted comes from a wide range of sources—censuses, surveys, sensors, commercial transactions, and records of all kinds. The information is made available not only by governmental entities, but also by businesses, the media, and other organizations, in the form of tables, graphs, maps, datasets, and other formats available today through the internet and other modes of access.

To meet the many and essential uses described above, information on a nation's society and economy must be credible and trustworthy. The consumers of the information must believe that the information is objective and not affected by any political or ideological perspective concerning the phenomena being measured. They must trust that the technical expertise of the producers of the information is sufficient to produce statistics that will meet their needs, which include consistency in definitions and methods so that one can judge whether things are getting better or worse over time and can compare different parts of the country.

Who produces such information to inform societal and economic planning, decision making, and knowledge generation and to power the myriad needs and requirements of modern democracies and advanced economies? Many actors provide useful information, but across the world statistical agencies in central governments have the role of producing key national statistics in ways that maximize their credibility and utility to inform policymakers and the public. Furthermore, by extension, statistical agencies also help citizens assess the reliability of the information released by other sources.

The United Nations General Assembly in 2014 formally endorsed *Fundamental Principles of Official Statistics* (see Appendix C).[1] The first of these principles accords worldwide recognition to the indispensable role of official statistics:

> Official statistics provide an indispensable element in the information system of a democratic society, serving the Government, the economy and the public with data about the economic, demographic, social and environmental situation. To this end, official statistics that meet the test of practical utility are to be compiled and made available on an impartial basis by official statistical agencies to honor citizens' entitlement to public information (United Nations Statistical Commission, 2014, pp. 1–2).

In this regard, national statistical information forms a data infrastructure that resembles the role of physical infrastructure for a nation, like interstate highways, national defense assets, interstate utility grids, and basic scientific research (see also American Statistical Association, 2024). All of these national investments serve the common good. Their benefits are sometimes relatively small for an individual and often unrecognized, but they are essential to the welfare of the whole country. In some sense, these infrastructures are among the threads of the fabric of society.

In their day-to-day lives, most people do not think about the benefits of highways and bridges—until they exhibit a problem, perhaps being closed for repairs, or worse, when they fail and collapse. So too, when statistical information is disrupted or compromised, its value is vividly illustrated by decisions that, in retrospect, appear misguided. As just one example, inadequate information that results in underestimating the depth of a looming recession or, conversely, an economic boom, can lead to less-than-optimal policies to rekindle or rein in growth (Reamer, 2014). As another example, national data on the societal impact of pandemics or environmental change, such as COVID-19 and wildfires, are essential for policymaking. Delays or incomplete national data miss opportunities to provide guidance to the public at a critical time. Issues such as these require ongoing monitoring of their impact across many facets of our society.

---

[1] These were originally adopted by the United Nations Statistical Commission in 1994; see Appendix C.

STATISTICAL INFORMATION POWERS POLICYMAKING

Following are a few examples of the many ways that statistical information provided by federal agencies serves the nation.

Informs Political Representation

The U.S. Constitution mandates a decennial census of the population every 10 years (the first census was taken in 1790) for determining the allocation of seats in the U.S. House of Representatives among the states. Reapportionment in turn triggers the redistricting process, by which states, using census data, redraw the boundaries of congressional districts to accommodate changes in the number of seats and in the geographic distribution of the population. States and many local governments also use census and other data to reapportion and redistrict their legislative bodies.

Informs Economic Decision Making

Federal statistics drive important decisions. The federal government currently labels 36 statistics—such as gross domestic product (GDP), the employment situation, monthly wholesale trade, weekly natural gas storage, crop production, consumer credit, and others—as "principal federal economic indicators."[2] The release of these statistics moves financial markets. The Office of Management and Budget's (OMB's) *Statistical Policy Directive No. 3* requires these indicators to be published by the cognizant statistical agency on specified release dates under procedures designed to protect the integrity and credibility of the estimates and ensure that they are not subject to manipulation and do not give any user an unfair advantage, so that businesses and the public can be confident the statistics are objective (OMB, 2024d). (See Appendix A, and Practice 2 in Chapter 3.)

The indicators often lead the headlines upon their release, contribute significantly to public- and private-sector decision making, and help inform the public as to where the nation has been and where it is going. They and many other federal social and economic statistics have real consequences: the Consumer Price Index (CPI) determines annual cost-of-living adjustments to Social Security and Supplemental Security Income monthly benefits, which in May 2024 amounted to $120.6 billion provided to 67.8 million people.[3] Annual changes in the CPI also affect commercial and residential rents, public- and private-sector wages, and components of the federal income tax code. Annual changes in prices for geographic areas

---

[2] See Statistical Policy Directive No. 3 in Appendix A.
[3] https://www.ssa.gov/policy/docs/quickfacts/stat_snapshot/

enter into local decisions, and monthly changes in prices are a major input into Federal Reserve Board decisions on short-term interest rates. Annual poverty estimates from the Small Area Income and Poverty Estimates Program of the Census Bureau are used to allocate funding to school districts under Title I of the Elementary and Secondary Education Act.[4]

Informs Business Decisions

The U.S. and global economies are powered by data.[5] Whether starting or expanding a business, exploring prospects for different occupations, anticipating demand for products, projecting the labor force, evaluating effects of trade patterns, targeting investments, forecasting energy prices, planning for hurricanes, funding pension plans, devising better ways to serve customers with disabilities, or finding suppliers, business owners and community members rely every day on data produced by the federal government. Beginning in the 1960s, data provided by the U.S. government electronically spurred the development of a new sector: firms[6] that provide government-data-related products to households, businesses, and organizations. Over the past 10 years, the revenue produced by industries that rely on government data has increased. Between 2012 and 2022, the revenues of government data-intensive sector (GDIS) grew from $407.9 billion to $778.0 billion. In this timeframe, the GDIS grew faster than the rest of the economy, increasing its share of U.S. Gross Domestic Product (GDP) by 50%, from 1.9% to 2.9%. In FY2023, the 13 principal statistical agencies—which is just part of the federal data collection—had a combined budget of roughly $3.5 billion.[7] There are numerous other uses by businesses and governments (see, e.g., Department of Commerce, 2014; Hughes-Cromwick & Coronado, 2019).

Assists Federal, State, and Local Government Action

Federal statistics provide high-quality, comparable information across the country. The American Community Survey (ACS), for example, provides key information that states and local governments use for disaster preparedness, economic development and workforce planning, public

---

[4]https://www.census.gov/topics/income-poverty/poverty/guidance/data-sources.html
[5]https://www.weforum.org/agenda/2016/01/a-global-economy-powered-by-data
[6]More recent examples of such firms include Google, KPMG, and Zilllow, to name just a few.
[7]See https://www.commerce.gov/news/blog/2024/12/revenue-industries-heavily-reliant-us-government-data. This paragraph and footnote were modified after release of the report to provide a more precise estimate of value and its context based on publication released on December 3, 2024.

health surveillance, and regional transportation planning (see National Academies of Sciences, Engineering, and Medicine [NASEM], 2019b; National Research Council [NRC], 2007a, 2013c). Data from the decennial census and the ACS are used to distribute hundreds of billions of dollars to states and localities for Medicaid, for education, and for housing, food assistance, veterans, and transportation programs.[8] Additional examples of federal data products that inform state-level analyses and decisions include: state- and substate-level monthly employment and unemployment statistics produced by the Bureau of Labor Statistics;[9] small-area estimates of poverty and population produced by the Census Bureau; state-level counts of deaths due to COVID-19[10] and drug overdose;[11] and state-level estimates of student performance based on data collected through the National Assessment of Educational Progress by the National Center for Education Statistics.

Monitors the Social and Economic Health of the Nation, States, and Localities

Regularly published social and economic indicators from statistical agencies are widely cited in the media and consulted by the public to identify trends and, when estimates are available for state and local areas, to compare across areas. Some examples include *America's Children: Key National Indicators of Well-Being* from the Interagency Forum on Child and Family Statistics; the *Condition of Education* from the National Center for Education Statistics; *Criminal Victimization* from the Bureau of Justice Statistics; *Statistics of Income* from the IRS; *Income and Poverty in the United States* from the Census Bureau; *Health United States* from the National Center for Health Statistics; and *Science and Engineering Indicators* from the National Science Board and National Center for Science and Engineering Statistics.[12]

---

[8] https://gwipp.gwu.edu/counting-dollars-2020-role-decennial-census-geographic-distribution-federal-funds#Briefs

[9] BLS produces these statistics using Current Population Survey and Unemployment Insurance data collected by the Census Bureau and individual states, respectively. In particular, the Local Area Unemployment Statistics program produces monthly and annual employment, unemployment, and labor force data for census regions and divisions, states, counties, metropolitan areas, and many cities, by place of residence. See https://www.bls.gov/lau/.

[10] https://www.cdc.gov/nchs/nvss/vsrr/COVID19/index.htm

[11] https://www.cdc.gov/nchs/nvss/vsrr/drug-overdose-data.htm

[12] https://www.childstats.gov/americaschildren/; https://nces.ed.gov/programs/coe/; https://www.bjs.gov/index.cfm?ty=tp&tid=9; https://www. irs.gov/statistics; https://www.census.gov/library/publications/2019/demo/p60-266.html; https://www.cdc.gov/nchs/hus/index.htm; and https://ncses.nsf.gov/pubs/nsb20201

Provides Evidence for Developing and Evaluating Public- and Private-Sector Programs

Data on the condition of housing and housing finance to inform housing policy come from the ongoing American Housing Survey (see NRC, 2008a). Statistics on the various types of energy used for heating, cooling, information technology, and other uses are provided by energy consumption surveys for commercial buildings and for residences (see NRC, 2012b). The Commission on Evidence-Based Policymaking highlighted a number of examples of how administrative data have been used to generate evidence to inform government policies, including substance abuse education and workforce investment (Commission on Evidence-Based Policymaking, 2017, pp. 9–10).

Provides Input to Social Science Research That Informs the Public

Many policy-relevant insights have resulted from statistical analysis of long-running federally funded surveys, including longitudinal surveys that follow individuals over time (see NRC, 2005). Some examples: the National Center for Education Statistics runs a number of longitudinal surveys following children through their K–12 education and post-secondary education and beyond to look at the transitions from high school and college to the labor force; the Bureau of Labor Statistics' National Longitudinal Surveys follow young adults through their working lives to look at career paths; and the Health and Retirement Study of the University of Michigan, supported by the National Institute on Aging and the Social Security Administration, follows older adults through retirement to look at health and well-being.[13]

THE COSTS AND BENEFITS OF FEDERAL STATISTICS

The cost of federal statistical programs is a tiny fraction of overall U.S. federal spending. In fiscal year (FY) 2022, the combined budget request for all the major statistical agencies and other statistical programs in federal agencies (including the decennial census, economic, and agricultural censuses) totaled $7.1 billion.[14] This amounted to about 0.03 percent of GDP[15]

---

[13] https://nces.ed.gov/surveys/; https://www.bls.gov/nls/; http://hrsonline.isr.umich.edu/

[14] See https://www.whitehouse.gov/wp-content/uploads/2024/02/statistical-programs-20212022.pdf. OMB includes all statistical programs with at least $3 million in estimated or direct funding in FY 2019, FY 2020, FY 2021, or FY 2022.

[15] https://www.bea.gov/news/2023/gross-domestic-product-fourth-quarter-and-year-2022-third-estimate-gdp-industry-and#:~:text=Imports%20increased%20(table%202).,(tables%201%20and%203)

or about 0.1 percent of the total appropriated budget of about $6.3 trillion for the federal government,[16] equivalent to about $21 per U.S. resident per year.[17] This tiny fraction of funds allocated for federal statistical programs has been declining in recent years,[18] despite ever-growing demands for more accessible and timely data (American Statistical Association, 2024).

The benefits from this investment in federal data collection and statistics permeate every corner of the United States. A recent publication from the Department of Commerce estimates the value of government data-intensive sector as $407.9 billion in 2012 rapidly growing to $778.0 billion in 2022.[19]

It is impossible to capture the full economic and societal value of having reliable data on economic, social, health, agricultural, industrial, and environmental characteristics of the country,[20] although some experts have estimated the annual value of making federal data from statistical and program agencies "open," that is, freely available to the public.[21] The United Nations (2018) report, *Recommendations for Promoting, Measuring, and Communicating the Value of Official Statistics*,[22] argued that official statistics have value far beyond their dollar worth. If the national government did not collect or acquire such data, the private sector might fill the breach—but likely at a greater cost to obtain data of comparable or less quality. Response rates would be lower compared with federal surveys, with no guarantee of continuance or continuity. The report went further to suggest the absence of government data collection and acquisition presented the possibility of a two-tier system whereby only those who could pay

---

[16] https://www.cbo.gov/publication/58888

[17] Based on a total U.S. resident population of 332,403,650. https://www.commerce.gov/news/blog/2022/01/us-population-estimated-332403650-jan-1-2022

[18] As an example, this decline in response rates can be seen in the National Crime Victimization Survey (declines in the person-level response rate from 80% in 2011 to 50% in 2021) and the National Health Interview Survey (declines in the adult response rate from about 70% in 2011 to 50% in 2021). See (American Statistical Association, 2024).

[19] See https://www.commerce.gov/news/blog/2024/12/revenue-industries-heavily-reliant-us-government-data. This paragraph and footnote were modified after release of the report to provide a more precise estimate of value and its context based on publication released on December 3, 2024.

[20] https://www.commerce.gov/sites/default/files/migrated/reports/the-value-of-the-acs.pdf tells how the ACS is used by federal, state, and local governments and by businesses, school districts, and academic researchers. A panel from the American Enterprise Institute discussed the immense value of government data for commerce, as well as the private companies that, essentially, repackage, build upon, and sell government data. See https://www.aei.org/wp-content/uploads/2017/02/170302-AEI-Vital-Role-of-Government-Statistics.pdf

[21] Making data collected by the federal government available to the public at no cost in a machine-readable format without restrictions on its use is referred to as "open data." See http://reports.opendataenterprise.org/2017OpenDataRT1-EconomicGrowth.pdf

[22] https://www.unece.org/fileadmin/DAM/stats/publications/2018/ECECESSTAT20182.pdf

would have access to the data they need. An additional cost is the potential loss of trust and credibility (and therefore, value) where private sector companies producing the data could have a stake in the results (for example, grain companies reporting on crop production; energy companies on gas production; and tobacco companies on the health effects of tobacco use).

The fundamental characteristic of federal statistics as a public good (see Box 2-1) and the demonstrated policy, planning, research, and informational value of today's portfolio of statistical programs justify adequate budgets for federal statistics. Such funding needs to provide for research and development for continuous improvement in relevance, accuracy, timeliness, and accessibility (see Practice 5). In turn, it is incumbent on federal statistical agencies to communicate the value of their programs to policymakers and others and to analyze the cost-effectiveness and value of their programs on a continuing basis so that they can ensure the best return possible on the tax dollars invested in them.

BOX 2-1
Federal Statistics as a Public Good

Familiar public goods, which benefit everyone in a country but cannot readily be provided through the marketplace, include the judiciary and the national defense. Formally, a public good has two components (see Gravelle & Rees, 2004). First, a public good must be nonrivalrous—that is, when one individual consumes it, there is no actual or potential reduction in the amount available for another individual to consume. A public good must also be nonexcludable, that is, it must be difficult for a business or other private entity to try to establish a market for such a good that is open only to those willing to pay the price. Federal statistics are essentially a public good—they satisfy the first component and practically speaking satisfy the second (NRC, 1999, Chapter 2).

In regard to the second component, a chief virtue of official federal statistics (e.g., the unemployment rate) is that they are designed to be uniquely representative of the whole country and/or the segments they cover. A private entity would face high costs and an inadequate return from providing such statistics on an ongoing basis. Consider the many examples of statistical information provided by businesses, academic institutions, and other nongovernmental organizations. The information may be based on a survey or another data source, and it may add value to an underlying federal data series. Yet nongovernmental organizations often do not attempt to gather information and provide timely, ongoing statistics on the scale necessary to represent the entire population of persons or organizations. Even the largest "Big Data" series often exclude the most vulnerable members

continued

BOX 2-1 Continued

of society who may not have cell phones or do not subscribe to or purchase a company's services.

Major federal household and business surveys and censuses provide vitally important information for a broad range of data users. Other federal statistical programs are important for scientific research and program evaluation (e.g., longitudinal surveys) and may have a specialized base of users. Both are costly to carry out. And, in either case, nongovernmental organizations are not likely to view them as viable business propositions. Business reasons may also lead a nongovernmental data holder to modify or abandon a useful statistical series.

Moreover, private-sector series often depend on federal statistics in various ways. For example, private-sector price indexes based on web-scraping (e.g., the MIT Billion Prices Index[a]) have used federal CPI information for market-basket weights and benchmarking, and public opinion polls and marketing surveys use federal statistics on gender, age, ethnicity, and other characteristics to adjust the raw data to represent population groups. Thus, federal statistics not only are useful and often indispensable themselves, but they also are a necessary ingredient to many other data products and services that fulfill information needs.

---

[a] https://thebillionpricesproject.com/

3

Principles

Highlights

- Five mutually reinforcing, co-equal principles guide federal statistical agencies:
  1. Relevance to Policy Issues and Society
  2. Credibility Among Data Users and Stakeholders
  3. Trust Among the Public and Data Subjects
  4. Independence from Political and Other Undue External Influence
  5. Continual Improvement and Innovation
- Each of these principles is supported by key practices.

PRINCIPLE 1: RELEVANCE TO POLICY ISSUES AND SOCIETY

Federal statistical agencies exist to provide accurate and timely information relevant for policy and public use. Federal statistics can describe economic conditions, delineate societal problems, and inform policymaking and the evaluation of programs. To provide the relevant statistical information needed by policymakers in Congress and the executive branch, as well as by other users, statistical agencies must have a solid understanding of the public policy issues, federal programs, and information needs in their domains and they must have a clear mission (see Practice 1). A major advantage of the decentralized federal statistical system in the United States is that separate federal statistical agencies are located in appropriate departments and are closer to the policymakers and programs in those areas. However, this can also present a challenge when different agencies produce different statistical estimates of the same or similar phenomena. It is essential that statistical agencies coordinate and collaborate with one another (see Practice 7) to ensure that coherent and consistent statistical information is provided on major policy issues, such as the federal collection and dissemination of race and ethnicity data (see Appendix A).

To ensure that they are providing relevant information, statistical agencies need to reach out to a wide range of their data users, including staff in their own departments and other federal departments who use their data, members of Congress and congressional staff, state and local government agencies, academic researchers, businesses and other organizations, organized constituent groups, and the media.

Agencies may need to expend considerable energy to open avenues of communication more broadly with current and potential users (see Practice 9). The recently introduced requirement that federal agencies create learning agendas could assist statistical agencies in prioritizing their outreach activities (Commission on Evidence-Based Policymaking, 2017; Foundations for Evidence-Based Policymaking Act of 2018, 2019; Office of Management and Budget [OMB], 2019b).

Statistical agencies have used a variety of approaches to engage with users. Advisory committees, such as the Census Bureau's National Advisory Committee, are one tool to obtain the views of users outside a statistical agency (National Research Council [NRC], 1993b, 2007a).[1] Many agencies obtain advice from committees that are chartered under the Federal Advisory Committee Act.[2] Some agencies obtain advice from committees and working groups that are organized by an independent association, such as the American Statistical Association's Committee on Energy Statistics for the Energy Information Administration. Regardless of the exact organization of these advisory committees, statistical agencies should examine the

---

[1] Some statistical agencies had advisory committees that were later disbanded by their departments. This has occurred for the Bureau of Transportation Statistics and the National Center for Education Statistics.

[2] Examples include: the Board of Scientific Counselors for the National Center for Health Statistics; the Data Users Advisory Committee and the Technical Advisory Committee for the Bureau of Labor Statistics; the Census Scientific Advisory Committee and the National Advisory Committee for the Census Bureau; and the Federal Economic Statistics Advisory Committee, which provides substantive and technical advice cutting across the major economic statistics programs of three agencies: the Bureau of Economic Analysis, the Bureau of Labor Statistics, and the Census Bureau. See https://apps.bea.gov/fesac/. See also PL-92-463.

institutional affiliation and occupation of advisory committee members to ensure equitable coverage in representation.

Additionally, holding workshops and conferences for data users or engaging with them at professional conferences are valuable activities for facilitating interchange among users and agency staff (NRC, 2013c). Online mechanisms, such as blogs and web surveys, may assist a statistical agency in obtaining input from users (see Practice 9). Similarly, agencies can use web analytics to better understand their user base and to assess the accessibility and usability of their website and data products. Offering positions to data users as fellows or temporary employees can also help a statistical agency gain a richer perspective on user interests and concerns.

Statistical agencies should periodically review their data collection programs and products to make sure they remain relevant (see Practice 6). Relevance should be assessed, not only for particular programs or closely related sets of programs but also for an agency's complete portfolio, to assist each agency in making the best choices among program priorities given available resources.

To increase data quality and relevance, an agency's own staff should actively analyze its data (Martin, 1981; Norwood, 1975; Triplett, 1991). Such analyses may examine correlates of key social or economic phenomena or study the statistical error properties of the data. Carrying out such work can lead to improvements in the quality of the statistics, to the identification of new needs for information and data products, to a reordering of priorities, and to a deeper understanding of data users' needs (see Practice 5).

The substantive analyses that statistical agencies produce as a regular part of their dissemination and research activities will likely be helpful to policy analysis units in their departments, as well as other data users. These analyses typically describe relevant conditions and trends over time and across geographic areas and population groups (e.g., high school completion rates by race, poverty rates for each year, or state variation in employment rates). A statistical agency may expand upon such initial analyses at the request of a policy analysis unit or other data user, for example by examining trends for particular population groups, while being mindful of the agency's responsibility to remain independent from undue influence (Principle 4). Further, statistical agencies have an obligation to provide useful statistical information to Evaluation Officers to conduct their work.

However, statistical agencies should be careful not to become involved with policy development or implementation (beyond policies directly affecting their operations), because these activities could affect their ability (or the perception of their ability) to conduct impartial and objective statistical activities. Examples of policies that are appropriate for statistical agencies to help develop and implement include statistical system policies, federal data quality standards, research access, privacy protections, and information technology security protocols—all as they relate to statistical operations (see Practices 2, 3, 4 and 8). Beyond such exceptions, a statistical agency should neither make policy recommendations nor conduct substantive analyses of policies, although it may advise on the availability and strengths and limitations of relevant information in a policy-neutral manner. The distinction between analysis consistent with the mission of a statistical agency and policy analysis is not always clear, and a statistical agency must carefully consider the extent of policy-related activities that are appropriate for it to undertake to maintain its primary mission of providing impartial and objective statistical information for public use (see Practices 1 and 2). Principle 1 is summarized in Box 3-1, below.

BOX 3-1
Summary: Principle 1 (Relevance)

Key Message:
Federal statistical agencies must provide objective, accurate, and timely information that is relevant to important public policy issues.

Key Supporting Practices:
Practice 1: A Clearly Defined and Well-Accepted Mission
Practice 2: Necessary Authority and Procedures to Protect Independence
Practice 5: An Active Research Program
Practice 6: Strong Internal and External Evaluation Processes for an Agency's Statistical Programs
Practice 7: Coordination and Collaboration with Other Agencies
Practice 9: Dissemination of Statistical Products That Meet Users' Needs

---

PRINCIPLE 2: CREDIBILITY AMONG DATA USERS AND STAKEHOLDERS

The value of a statistical agency rests fundamentally on the accuracy and credibility of its data products. Because few data users have the resources to verify the accuracy of statistical information, users rely on an agency's reputation to disseminate high-quality, objective, and useful statistics in an impartial manner. Only if its products are viewed as credible can an agency be regarded as working in the national interest, not beholden to a particular set of users (Ryten, 1990; see Practice 2).

Credibility therefore stems from the respect and trust of users and stakeholders in the statistical agency. Agencies build this respect and trust, not only by producing accurate and objective data and meeting all of their deadlines for the release of their statistics, but also by adhering to the other principles for federal statistical agencies and by following some key practices. When different agencies produce similar or related estimates, it is essential that statistical agencies coordinate and collaborate with one another (see Practice 7) to ensure that users understand the differences and can determine which data are most useful for their needs.

Agencies build and maintain respect and trust through clear public commitments to professional practice and transparency in all that they do. For example, statistical agencies should actively engage with users in determining priorities for data collection and analysis; make their data widely available on an equal basis to all users, formally and informally (see Practice 9); conduct research to improve efficiency and data quality (see Practices 3 and 5); and fully inform users of the strengths and limitations of the data (see Practice 10). Such activities demonstrate an agency's respect toward, and openness with, its users and stakeholders.

A statistical agency's website is a key vehicle for conveying not only its statistical data, but also key information about its data. Providing clear and easy access for users to locate, work with, and understand the strengths and limitations of the agency's data is a vital part of an agency's mission (see Practice 1) and requires ongoing efforts to continue to meet users' and stakeholders' evolving needs. An agency's website can enhance its credibility by providing information about its policies for data access (e.g., explaining which tables and microdata files are publicly available and which data require approval to access in secure sites to protect confidentiality; see Practice 8); scientific integrity policies; standards for data quality and for documenting sources of error in data collections and estimation models (see Practice 3); procedures and schedules for the release of new and continuing data series; procedures for timely notice of errors and corrections to previously released data; procedures and schedules for archiving historical data; and documentation of ongoing research efforts to provide accurate statistics that meet users' needs (see Practice 5).To support its credibility, it is essential that each statistical agency website be clearly distinguishable from its associated parent agency, particularly if that parent agency is a regulatory, direct service, or policy development agency.[3]

To keep up with an ever-changing society and technology and thereby maintain the trust of their users and stakeholders, statistical agencies need to recruit, develop, and retain high-quality professional staff who are dedicated to providing high-quality products and upholding high ethical standards (see Practice 4). Statistical agencies also need to regularly review and evaluate

---

[3]Note that the Trust Regulation authorizes each recognized statistical agency or unit to maintain a distinctive, outward-facing website with its own domain name and with adequate control over the website content and management to uphold the fundamental responsibilities. See final § 1321.4(e)(3) and Appendix A.

their programs and share the results of these evaluations with their users and stakeholders (see Practice 6). Principle 2 is summarized in Box 3-2, below.

BOX 3-2
Summary: Principle 2 (Credibility)

Key Message:
Federal statistical agencies must have credibility with those who use their data and information.

Key Supporting Practices:
Practice 1: A Clearly Defined and Well-Accepted Mission
Practice 2: Necessary Authority and Procedures to Protect Independence
Practice 3: Commitment to Quality and Professional Standards of Practice
Practice 4: Professional Advancement of Staff
Practice 5: An Active Research Program
Practice 6: Strong Internal and External Evaluation Processes for an Agency's Statistical Programs
Practice 7: Coordination and Collaboration with Other Agencies
Practice 8: Respect for Data Subjects and Data Holders and Protection of Their Data
Practice 9: Dissemination of Statistical Products That Meet Users' Needs
Practice 10: Openness About Sources and Limitations of the Data Provided

---

PRINCIPLE 3: TRUST AMONG THE PUBLIC AND DATA SUBJECTS

Nearly every day of the year, individuals, household members, businesses, state and local governments, and other organizations provide information about themselves when requested by federal statistical agencies. Without the cooperation of these data subjects, federal statistical agencies could produce very little useful statistical information.

Some information provided is required by law or regulation for government tax and transfer programs, such as reports of employers' wages to state employment security agencies or payments to program beneficiaries. A small number of federal statistical surveys are so important that participation is mandatory. But most of the data come from the voluntary cooperation of respondents. In all cases, the willing cooperation of data subjects reduces costs and promotes accuracy.

Because virtually every person, household, business, state or local government, and organization is the subject of some federal statistics, public trust is essential for the continued effectiveness of federal statistical agencies. Individuals and entities providing data directly or indirectly to federal statistical agencies must trust that the agencies will appropriately handle and protect their information. Implicitly and explicitly, they expect the following:

1. The agency's computer systems will not be hacked and their information taken;
2. Their information will not be accessed and used by others for non-statistical purposes;
3. The statistical agency has a lawful, legitimate need for each piece of information it is requesting;
4. The statistical agency will use their information to create useful statistics;
5. The statistical agency will not release or publish their information in identifiable form;
6. The statistical agency will control who is allowed to see and access their information; and
7. The statistical agency will evaluate and update its systems for using and protecting data subjects' information to take account of new information needs, new threats to confidentiality, and new technological developments that offer improved protection.

Federal statistical agencies not only have legal and ethical obligations that require them to fulfill these expectations; they also have the obligation to effectively communicate how they fulfill them. Consistent ethical conduct on the part of a statistical agency is critical for obtaining the trust of the general public and of data subjects, whether those subjects are individuals, organizational entities, or custodians of administrative records. Statistical agencies should coordinate and collaborate with each other to ensure that their communications and internal practices are clear and consistent, as this will strengthen the trust of data subjects who may interact with more than one agency (see Practice 7).

Data subjects must trust that the information the agency seeks is important for the government to collect and is being collected in a competent manner, for the good of the larger society, and only for the purposes that the agency has described (see Practice 2). To engender trust, a statistical agency should also respect the privacy of data subjects in other ways and ensure that each individual's consent to respond to a survey is given knowingly and with full information. Agencies should describe the intended and likely future uses of the data being collected, the data's relevance for important public purposes, and the extent of confidentiality protection that will be provided. Agencies should minimize the intrusiveness of questions and the effort needed to respond to them, and they should seek administrative or other non-survey sources to fulfill needs consistent with each agency's requirements for information (see Practices 5 and 8). Trust among data subjects also requires that an agency treat respondents with courtesy in appreciation for their time (National Academies of Sciences, Engineering, and Medicine [NASEM], 2016a).

The mission of federal statistical agencies is to produce statistical information by aggregating the data provided by individuals, businesses, or other entities. These agencies pledge to use the information they collect only for statistical purposes, not to provide individual records for any administrative, regulatory, or judicial use, and to make every effort to protect the confidentiality of individual information in the data they publish. This pledge is supported by many statistical agencies' individual statutes as well as the Confidential Information Protection and Statistical Efficiency Act of 2018 (Title III of Foundations for Evidence-Based Policymaking Act of 2018, 2019). (See Appendix A.)

Data subjects must be able to trust that a statistical agency will scrupulously honor its pledge of confidentiality and will fulfill the expectations noted above. Earning this trust, however, goes beyond what the agency is simply required to do by law, and recognizes that there are many potential threats, some outside the control of the statistical agency, that the agency must anticipate and guard against. In the world of "big data" and the "dark web," agencies must guard against the use of external data to re-identify information provided by individuals. Agencies must consistently innovate in privacy-protecting technologies to protect—to the extent possible—against re-identification of individual records in statistical data products. (See NASEM, 2017a,c, 2024c; and Practice 8.)

When data are obtained from the administrative records of other federal, state, or local government agencies or other third-party providers, the same requirements of trust building apply to justify their cooperation.[4] Data-providing organizations need to trust that their records are important and legitimate for a statistical agency to obtain, that their own restrictions on data access will be honored, and that the statistical agency will make every effort to minimize their burden in responding.

An effective statistical agency has policies and practices to instill the highest possible commitment to professional ethics among its staff and build a culture of confidentiality. When issues arise or guidance is unclear, it must be able to rely on its staff to keep this culture (see Practices 3 and 4).

We end by acknowledging that statistical agencies' mandate to establish and maintain the trust of data subjects, holders, and users is made more difficult when overall trust in government data and government agencies

---

[4] See (OMB, 2014a) in Appendix A, which asserts the legitimacy and benefits of use of administrative data from other federal agencies for statistical agency purposes. It also provides guidance for best practices and procedures to engender mutual respect and trust and facilitate such data sharing.

is waning (American Statistical Association, 2024; NASEM, 2023b). Some of this diminished trust is due to disparagement of particular statistical releases or agencies by public figures. An unfavorable climate raises the need for statistical agencies to focus intently on practices that help to build and maintain trust. Principle 3 is summarized in Box 3-3, below.

BOX 3-3
Summary: Principle 3 (Trust)

Key Message:
Federal statistical agencies must have the trust of those whose information they obtain.

Key Supporting Practices:
Practice 2: Necessary Authority and Procedures to Protect Independence
Practice 3: Commitment to Quality and Professional Standards of Practice
Practice 4: Professional Advancement of Staff
Practice 5: An Active Research Program
Practice 7: Coordination and Collaboration with Other Agencies
Practice 8: Respect for Data Subjects and Data Holders and Protection of Their Data

---

PRINCIPLE 4: INDEPENDENCE FROM POLITICAL AND OTHER UNDUE EXTERNAL INFLUENCE

A statistical agency must be impartial and execute its mission without being subject to pressures to advance any political or personal agenda. It must avoid even the appearance that its collection, analysis, or reporting processes might be manipulated for political or other purposes. Only in this way can a statistical agency serve as a trustworthy source of objective, relevant, accurate, and timely information (Bohman, 2024; Bowen, 2023; Citro et al., 2023; Cohen, 2023; Habermann & Louis, 2020; Habermann et al., 2023).[5] Their independence and their high-quality products also allow statistical agencies to perform another increasingly important service for the country: they assist and improve citizens' understanding of statistical products amid an avalanche of information of varied provenance and quality (American Statistical Association, 2024; Behzad et al., 2023).

---

[5] The importance of independence from undue external influence is reflected in national statute and regulation, as well as international guidance. See (Foundations for Evidence-Based Policymaking Act of 2018, 2019; OMB, 2014b) in Appendix A and (European Statistical System Committee, 2017; United Nations Statistical Commission, 2014) in Appendix C.

Statistical agencies and the statistical data they produce can play a key role in *informing* policymakers, but they are not and should not be responsible for *developing* or *implementing* policy (see Practice 1) beyond policies relevant to the activities of the statistical agencies themselves (as noted in Principle 1). For this reason, statistical agencies should be distinct from units of a larger department that carry out administrative, regulatory, law enforcement, or policymaking activities.

It is also essential that a statistical agency be independent of other undue external influence in developing, producing, and disseminating statistics. "Undue external influences" are those from outside the agency that seek to undermine its impartiality, nonpartisanship, or professional judgment. Independence from any undue outside influence fosters trust among data subjects and credibility with data users. Examples of undue external influence include attempting to dictate measurement methods for a statistical agency's programs; to delay or speed up (e.g., before or after an election) the release of statistical data; to suppress or alter scientific or technological findings and conclusions; and to alter the content of press releases to "spin" the findings to promote a particular viewpoint. Undue external influence also includes creating new positions for political appointees in a statistical agency with decision authority over the content, methods, and release of a statistical agency's data.

To fulfill this principle, a statistical agency must have the necessary authority and support to protect its independence (see Practice 2); however, a broad view of this authority is needed. Statistical agencies exist in a complex ecosystem and are governed by their legislative authority, which may give ultimate responsibility for the activities of the agency to the secretary of the department, as well as by the OMB and congressional committees. Within this broad framework, a statistical agency must maintain its credibility as an impartial purveyor of information (American Statistical Association, 2024; Bowen, 2023; Citro et al., 2023; Cohen, 2023; Habermann & Louis, 2020; Habermann et al., 2023). The provisions of a statistical agency's authorizing legislation can help promote its independence from political or other undue external influences.

For the head of an agency, independence and protection from undue political influence can be strengthened by the method of the person's appointment (Habermann et al., 2023). A method widely regarded as bolstering the professional independence of an agency head is appointment by the President with confirmation by the Senate for a fixed term and with a statutory requirement that the appointee be selected with appropriate professional qualifications, as is the case for the commissioner of the Bureau of Labor Statistics (BLS) and the director of the Census Bureau.[6] It may also be desirable that

---

[6]The heads of the Bureau of Justice Statistics (BJS), the Bureau of Transportation Statistics (BTS), and the National Center for Education Statistics (NCES) had been appointed by the

the leader's term not coincide with the presidential term to better ensure that professional criteria, rather than political ones, guide the appointment process. Appointment by the President with Senate confirmation for a term that is at the pleasure of the President, as is the case for the head of the Energy Information Administration (EIA), provides less assurance of independence (however, it is worth noting that EIA does have other strong legislative protection for the authority of its administrator). Appointment of a qualified career civil servant as the head of an agency is another method considered helpful for maintaining the independence of a statistical agency.[7]

Having its agency head report directly to the secretary of the department can also be helpful for a statistical agency to maintain a position of independence from political or other undue external influence. Such access allows the head to inform new secretaries about the role of their statistical agency and to directly present the case for new or changed statistical initiatives. Such direct access currently is provided by legislation for BLS and EIA. Other statistical agencies have one or more layers of departmental management between the statistical agency head and the secretary (see Figure B-2 in Appendix B). Over time there has been an increase in the "layering of statistical agencies," that is, positioning them lower in their department's administrative structure, a trend that the National Research Council (NRC, 2009c, p. 226) has identified as "a subtle, but increasingly common" threat to independence because it increases the number of political appointees and career staff who could seek to exercise control over the agency without transparency to external stakeholders or users (Habermann et al., 2023; Hartman et al., 2014).

Another means to protect against political and other undue external interference is for the statistical agency to have its own funding appropriation from Congress separate from that for other departmental agencies or programs.[8] (See Practice 2.) This provides greater visibility and accountability to Congress, both by the agency and by its department, something

---

President and confirmed by the Senate in the past; these positions were changed, two to become Presidential appointments without Senate confirmation (BJS and NCES) and one to become a career civil servant (BTS).

[7]Agencies headed by career civil servants, many of whom hold their positions for long periods of time, include the Bureau of Economic Analysis; the Bureau of Transportation Statistics; the Economic Research Service in the U.S. Department of Agriculture; the National Agricultural Statistics Service; the National Center for Health Statistics; the National Center for Science and Engineering Statistics (NCSES); the Office of Research, Evaluation, and Statistics (ORES) in the Social Security Administration; and the Statistics of Income (SOI) Division in the Internal Revenue Service.

[8]Most, but not all, federal statistical agencies receive appropriations directly. Three federal statistical agencies—National Center for Science and Engineering Statistics, Statistics of Income, and Office of Research, Evaluation, and Statistics—receive funding through appropriations to their parent agencies. In addition, some federal statistical agencies, such as NCES, are not involved in the annual budget request and communication process.

that is reinforced when the statistical agency participates in appropriations briefings. Other funding arrangements, such as the statistical agency being completely dependent on allocations from the budget of its parent department or agency, risk giving the department a great deal of leverage over the statistical agency without transparency to external stakeholders and users, and potentially compromising its ability to fulfill its mission.

A key aspect of a federal statistical agency's mission is its ability to release its statistical products without review or approval by policy officials outside the statistical agency (see Practice 2). Some agencies have this authority spelled out in statute, while others have departmental policies that support it. OMB provides governmentwide protocols and assurances on the release of key federal statistics and publishes in advance a release calendar for the entire year for Principal Federal Economic Indicators.[9] A strong internal and external evaluation program (see Practice 6) can also help ensure that all agency statistical programs are adhering to standard procedures and are not manipulated.

In the long run, the effectiveness of an agency depends on its reputation for impartiality: thus, an agency must be continually alert to possible infringements on its credibility and be prepared to strenuously resist such infringements (Bohman, 2024). OMB has stated that it is also the responsibility of the statistical agency's department to "enable, support, and facilitate federal statistical agencies and recognized statistical units" as they implement their responsibilities to produce objective data.[10] A federal statistical agency that has the respect and trust of its stakeholders and users, who can help publicly defend the agency, is better equipped to ward off or resist attempts by others to exert political or other undue external influence on the agency (see Principle 3). Within an agency, the professional staff's adherence to the mission of the agency and its quality standards and ethical principles (see Practices 3 and 4) is also key to the agency's preservation of its independence from political or other undue external interference.

Over the past decade, the federal government has issued a series of policies and frameworks to reaffirm and further support scientific integrity, including statistical production. These policies aim to protect against undue external influence in scientific processes by specifically rejecting the suppression or alteration of scientific findings by political officials; requiring transparency and accessibility of scientific findings; and requiring the selection of policy staff in scientific roles to be based on scientific and technological knowledge, credentials, experience, and integrity. (See Appendix A.) Principle 4 is summarized in Box 3-4, below.

---

[9]See OMB's Statistical Policy Directives No. 3 (OMB, 1985) and No. 4 (OMB, 2008) in Appendix A.

[10]See OMB's Statistical Policy Directive No. 1 (OMB, 2014b) in Appendix A.

> **BOX 3-4**
> **Summary: Principle 4 (Independence)**
>
> **Key Message:**
> Federal statistical agencies must be independent from political and other undue external influence in developing, producing, and disseminating statistics.
>
> **Key Supporting Practices:**
> Practice 1: A Clearly Defined and Well-Accepted Mission
> Practice 2: Necessary Authority and Procedures to Protect Independence
> Practice 3: Commitment to Quality and Professional Standards of Practice
> Practice 4: Professional Advancement of Staff
> Practice 6: Strong Internal and External Evaluation Processes for an Agency's Statistical Programs

PRINCIPLE 5: CONTINUAL IMPROVEMENT AND INNOVATION

Federal statistical agencies cannot be static. They must continually improve and innovate to be able to create reliable information on new policy questions, to provide objective information in a cost-effective way, to reduce respondent burden, and to meet user demands for more timely and granular information for statistical purposes.

Policy needs shift and evolve, and the society and economy that federal statistical agencies seek to measure are also evolving and changing at a rapid pace. To provide relevant information, statistical agencies must attend to changes in policy issues in their domain, identify emerging needs, and work with their data users and stakeholders to identify gaps in the agency portfolio or potential new statistical products that are needed (see Practice 9; NASEM, 2020a, 2023e; OMB, 2024b). One option to address needs for new information is for the agency to create experimental series; doing so allows the agency and its users time to evaluate a new data product without impacting existing data series (see Practice 5; NASEM, 2021b). The past few decades have seen an explosion of new data sources, some providing more geographic detail and timelier (some in near real time) information than federal statistical programs. Users have come to expect more, better, and ever more timely data. At the same time, individuals and businesses have been less and less willing to complete federal surveys and provide information to the government (a phenomenon that also affects private-sector surveys). Declining response rates have increased agency data collection costs, while federal statistical agency budgets have generally declined in real terms for more than a decade. Thus, agencies need to improve and innovate even to maintain their current programs.

These tensions of increasing user demand for relevant and timely statistics, agency requirements for credible data products, and significantly declining response rates amid increasing costs result in a critical challenge for statistical agencies today. The issues involved have been well-established (Advisory Committee on Data for Evidence Building, 2022; American Statistical Association, 2024; Commission on Evidence-Based Policymaking, 2017; NASEM, 2021a), but solving them requires statistical agencies to develop creative, innovative, and collaborative solutions to maintain the integrity of the statistics that are produced. Recent innovations in data sharing policies (see Practice 9 and Appendix A) appear promising to deliver improved access to data users and enhanced leveraging of alternative data sources to offset survey data collections, where possible. Additionally, examining and signaling sufficient fitness for use for a given statistical product could help agencies manage the level of effort required for data quality. These initiatives must be monitored rigorously.

Agencies should pursue state-of-the-art methods to acquire data, produce statistics, and provide access to their underlying data. Recent federal policy emphasis on open data sources and open data tools, most notably in the OPEN Government Data Act, Title II of the Evidence Act, and implemented in part through Data.gov, has promoted wider access to data for evidence building (Foundations for Evidence-Based Policymaking Act of 2018, 2019; OMB, 2013b, 2016b). Title II requires federal agencies to publish their nonconfidential data files online as open data, using standardized, machine-readable data formats, with their metadata included in the Data.gov catalog.

Agencies should engage in regular, periodic reviews of their major data collection programs that consider how to produce relevant, accurate, and timely data in the most cost-effective manner possible, while seeking to maintain comparability in key statistics over time and across geographies (see Practice 6). In ongoing programs that would be disrupted by the implementation of continuous improvements, a common practice is to bundle changes to implement several at the same time. For example, classifications such as the North American Industry Classification System (NAICS) are updated every 5 years and agencies may implement other changes at the same time as these updates occur. Agencies should ensure that the intervals between major research and development activities do not become so long that data collection programs deteriorate in quality, relevance, and efficiency (see Practice 6). When changes are made to ongoing data series, agencies should provide information to help users bridge across the old and new series.

Effective evaluations and communication are particularly necessary to inform and implement agencies' decisions to terminate programs or series (see Practice 6). Such terminations can be an essential step to allow an agency to maintain relevance when its resources are constrained.

An effective statistical agency keeps up to date on developments in theory and practice that may be relevant to its program. Examples of such developments include new uses for data about collection processes (that is, paradata); new techniques for imputing missing data (NRC, 2004b, 2010a) or for combining data from more than one source and estimating error in the resulting statistics; new methodologies for addressing data confidentiality and disclosure avoidance; and new techniques, such as artificial intelligence, for analyzing and processing data (NASEM, 2017a,d, 2023c; NRC, 2013a; OMB, 2024b).

Among several new developments in statistical methods, perhaps the greatest interest has been in the advancement of artificial intelligence and its growing application to both data production and analysis. Artificial intelligence methods are now being applied to statistical processes, for example to improve automated coding of job titles to standard occupation classification categories. Nonetheless, like the whole of society, federal statistical agencies are challenged to consider ways to use artificial intelligence techniques to improve timeliness and efficiencies while accounting for equity and quality concerns (National Artificial Intelligence Initiative Act, 2020; OMB, 2024f; Office of Science and Technology Policy, 2022). Innovation in this area will require sufficient staff and information technology resources to maintain existing programs while experimental data and methods are tested, reviewed, improved, and subsequently adopted (Executive Office of the President, 2023; Office of Science and Technology Policy, 2023a).

Statistical agencies need a robust research program that includes statistical methods, quality assessments, and evaluations of potential new data sources.[11] An effective statistical agency seeks out and carefully evaluates the quality and utility of potential new data sources and methods to harness information that could be useful for statistical purposes. Nontraditional data sources, such as sensor or transactions data, and fuller use of administrative records can potentially contribute to statistical programs by (a) augmenting information obtained from traditional sources such as surveys; (b) replacing information elements previously obtained from traditional sources; (c) providing earlier estimates that are later benchmarked with traditional sources; and (d) analyzing information streams to identify needed changes (see Practice 5). Agencies also need the appropriate information technology infrastructure and technical skills to handle alternative data sources. History has repeatedly shown that research conducted within federal statistical agencies on subject areas, methods, and operations can lead to large productivity gains in statistical activities for a relatively low cost (Citro, 2016; NRC, 2010c).

---

[11] See for example https://www.statspolicy.gov/assets/docs/ICSP-The%20Use%20of%20Private%20Datasets%20by%20Federal%20Statistical%20Programs-1-6-2023.pdf.

An effective statistical agency has a culture of continual improvement and innovation. All employees, and not just research staff, should be encouraged to seek to innovate and improve their functions within the organization. Staff in production and support areas should seek to improve processes, methods, and cost-effectiveness (see Practice 3). A statistical agency also needs to hire staff with cutting-edge skills and maintain and enhance the skills of its current staff through ongoing training and development opportunities so that it can continually improve and innovate (see Practice 4). To take the greatest advantage of staff with new and improved skills and to better support their operations, statistical agencies should maintain and regularly upgrade their information technology infrastructure (NASEM, 2017a, 2023c).

The decentralized nature of the U.S. federal statistical system can make it difficult for federal statistical agencies to easily learn from each other, but interagency and international collaborations can provide important and useful means for improving statistical programs. Some issues, such as accessing and using new data sources (NRC, 2008a), are common to many statistical agencies and can benefit from collaborative research across organizations (see Practice 7).

Finally, it is important to note that the imperative to innovate implies that successful statistical agencies must be able to devote sufficient resources to many components of sustaining innovation on an ongoing basis. That is, agencies must be able to modernize and develop new products even as they continue reporting on a timely basis within their existing systems. Resource availability for innovation is now complicated by the rising costs of conducting existing surveys, due to falling response rates. Principle 5 is summarized in Box 3-5, below.

BOX 3-5
Summary: Principle 5 (Innovation)

Key Message:
Federal statistical agencies must continually seek to improve and innovate their processes, methods, and statistical products to better measure an ever-changing world.

Key Supporting Practices:
    Practice 3: Commitment to Quality and Professional Standards of Practice
    Practice 4: Professional Advancement of Staff
    Practice 5: An Active Research Program
    Practice 6: Strong Internal and External Evaluation Processes for an Agency's Statistical Programs
    Practice 7: Coordination and Collaboration with Other Agencies
    Practice 9: Dissemination of Statistical Products That Meet Users' Needs

4

Practices

Highlights

- Ten practices support the achievement of principles guiding federal statistical agencies:
  1. A Clearly Defined and Well-Accepted Mission
  2. Necessary Authority and Procedures to Protect Independence
  3. Commitment to Quality and Professional Standards of Practice
  4. Professional Advancement of Staff
  5. An Active Research Program
  6. Strong Internal and External Evaluation Processes for an Agency's Statistical Programs
  7. Coordination and Collaboration with Other Agencies
  8. Respect for Data Subjects and Data Holders and Protection of Their Data
  9. Dissemination of Statistical Products That Meet Users' Needs
  10. Openness About Sources and Limitations of the Data Provided

- These practices require policy support, funding, and staff resources to implement.

PRACTICE 1: A CLEARLY DEFINED AND WELL-ACCEPTED MISSION

A statistical agency's mission should serve as a foundation not only for the work it does but also for how it does its work. Some agency missions are clearly spelled out in legislation; other agencies have only general authority granted them by legislation. Sometimes specific requirements are set by legislation or regulation (see federal-wide requirements for agency strategic plans and evaluation plans, such as in Title I of the Evidence Act, in Appendix A; see agency descriptions in Appendix B). A statistical agency's mission includes its responsibility to:

1. Produce and disseminate relevant and timely statistical information;
2. Conduct credible and accurate statistical activities;
3. Conduct objective statistical activities; and
4. Protect the trust of information providers by ensuring the confidentiality and exclusive statistical use of their responses.[1]

These responsibilities should be so ingrained into agency staff during their training and through the procedures and practices they follow that they become part of the culture of the agency. To be effective, a statistical agency also should

1. Ensure the quality of all aspects of its statistical programs, including measurement methods, data collection and processing, and appropriate methods of data analysis;
2. Evaluate, implement, and document new methods and processes that better serve users' needs (see Practice 5);
3. Curate its data to ensure their availability for future use, as well as documenting the methods used and the quality of the estimates (see Practice 9); and
4. Train its staff in a culture of responsible statistical practice.

Because nonstatistical activities threaten public trust in the agency, a statistical agency's mission must focus on information that is to be used for statistical purposes. A statistical agency should defend its mission and resist external attempts to extend its work beyond statistical purposes (see Practice 2). If a statistical agency is charged with collecting information for nonstatistical purposes (e.g., collecting data, not only for statistical purposes, but also for possible use in administrative actions affecting an

---

[1] See 44 USC § 3563(a)(1); originally issued as *Statistical Policy Directive No. 1* (see Appendix A).

individual), the agency should carefully segregate the statistical activities from the nonstatistical ones (e.g., perhaps locating the latter within a clearly demarcated office). If the senior leadership of the agency conclude that it is not possible to develop a satisfactory arrangement responsive to the agency's statistical mission, they should request that the activity be assigned elsewhere. Departments with federal statistical agencies have the responsibility to support and facilitate those statistical agencies in carrying out their mission and should not impose nonstatistical activities on them.[2]

Title III of the Foundations for Evidence-Based Policymaking Act of 2018 (2019; Evidence Act), also known as CIPSEA 2018, effectively expanded the mission of the federal statistical agencies with new authorities, roles, and responsibilities for evidence-based decision making. Statistical agencies should expand their administrative and alternative data holdings to develop evidence, as well as facilitate and expand secure, privacy-protected data access for evidence-building purposes. Statistical agencies should strategically implement these new authorities and responsibilities in order to maximize their mission impact (National Academies of Sciences, Engineering, and Medicine [NASEM], 2022a). Also, under the Evidence Act, statistical agency heads have the responsibility of serving as their cabinet department's chief Statistical Official[3] to lead statistical policy and activities, including setting standards for data quality and confidentiality. The Statistical Official should work closely with other senior officials, such as the Chief Data Officer and Chief Evaluation Officer, and other bureaus to advance the development and use of scientifically rigorous evidence, as well as promulgating good statistical principles and practices throughout the department (see Practice 7 and Appendix A).

A statistical agency should publicly communicate its mission and disseminate its statistical information and associated documentation on its website and other appropriate venues. The website should also provide information about enabling legislation, the scope of the agency's statistical programs, confidentiality provisions, data quality guidelines, and data access procedures. Consequently, agencies should carefully design their websites to maximize their utility to their users, stakeholders, and the public.

A statistical agency should periodically review its mission. As part of strategic planning to carry out its mission within its budget, it should review priorities among different programs, the infrastructure (e.g., computing capabilities, staff with appropriate expertise) needed to support them, and the relative urgency of needed improvements, say, in timeliness versus

---

[2] See 44 USC § 3563(b).

[3] In the case of the Departments of Agriculture, Commerce, and Health and Human Services, which each host more than one recognized statistical agency or unit, the Statistical Official role is determined by the Chief Financial Officer (CFO) Act agency.

accuracy. Statistical agencies should regularly evaluate their programs to determine whether they are fulfilling the agency's mission (see Practice 6), and an agency may need to eliminate or cut back an existing program in favor of a new initiative to better meet its mission (see Practices 5 and 9).

PRACTICE 2: NECESSARY AUTHORITY AND PROCEDURES TO PROTECT INDEPENDENCE

To maintain its credibility and reputation for providing objective, relevant, and accurate information, a federal statistical agency must have authority to maintain its independence from political and other undue external influences. Within an agency's ecosystem—which includes its own department, the Office of Management and Budget (OMB), and Congress—there are often important safeguards for its independence. In some cases, these are enshrined in law, such as the requirement that data collected for exclusively statistical purposes may not be used for law enforcement.[4] Other safeguards exist as longstanding governmentwide directives promulgated by the Office of the Chief Statistician in OMB that, for example, specify strict procedures for the release of statistical information that moves financial markets.[5] Others may exist as departmental policies or agency policies, widely accepted norms, or longstanding practices.

Some statistical agencies have more safeguards for their independence built into their originating statutes than others do,[6] while others rely on a history of having certain authorities without formal acknowledgement by their department. The proper functioning of individual agencies and the entire federal statistical system requires that there be strong and uniform recognition that these agencies have the authority to do the following:

1. Make decisions over the scope, content, and frequency of data compiled, analyzed, and disseminated within the agency's authorizing statutes based on sound scientific and professional considerations;
2. Select and promote professional, technical, and operational staff based on their professional skills and knowledge (see Practice 4);
3. Release statistical information, including accompanying press releases and documentation, without prior clearance regarding the statistical content of the release;[7]

---

[4] See Confidential Information Protection and Statistical Efficiency Act of 2018 (2018), Appendix A.
[5] See OMB *Statistical Policy Directive No. 3* (OMB, 1985), Appendix A.
[6] For example, the statute creating the Energy Information Administration specifically gives the Administrator the right to release statistical information without review by the Department of Energy.
[7] See *Statistical Policy Directives No. 3* (OMB, 1985) and 4 (OMB, 2008) in Appendix A.

4. Be able to make pledges to data subjects and other data holders that their data will be kept confidential and used only for statistical purposes (see Practice 8); and
5. Be able to meet with members of Congress, congressional staff, and the public to discuss the agency's statistics, resources, and staffing levels.

In order to provide objective statistical information, a statistical agency must have highly qualified staff (see Practice 3), who can make decisions on the scope, content, and frequency of data compiled, analyzed, and disseminated without political or other undue external influence. Their decisions should be based solely on scientific and professional considerations. These decisions should be well informed by consultations with users and stakeholders, including policy officials in their department, on their need for information (see Practices 5 and 9), and they must also meet statutory requirements for content and OMB clearance of information collections.

The selection of qualified professional staff, including senior executive career staff, should be determined by the statistical agency. While departments may need to approve some appointments, they should allow great discretion to the statistical agency in selecting staff with appropriate expertise. Agency staff who report directly to the agency head should have formal education and deep experience in the substantive, methodological, operational, and management issues facing the agency, as appropriate for their positions. For the head of a statistical agency, professional qualifications are of the utmost importance, whether the profession is that of statistician or is in a relevant subject-matter field (National Research Council [NRC], 1997a). Relevant professional associations can provide valuable input about suitable candidates.

Statistical agencies must protect the confidentiality of the data they acquire throughout the lifecycle of those data and their use.[8] Thus, statistical agencies must be able to exercise appropriate control over their data and the information technology (IT) systems on which they reside to securely maintain the integrity and confidentiality of individual records, ensure that the data can only be used for statistical purposes, and reliably support timely and accurate production of key statistics. A statistical agency must demonstrate the integrity, confidentiality, and impartiality of the data collected and sustained under its authority to maintain the trust of its data subjects, data holders, and data users (see Practices 8 and 9). Such trust is fostered when a statistical agency has control over its IT resources and there is no opportunity or perception that policy, program, or regulatory agencies could gain access to records of individual respondents. When departments

---

[8] See CIPSEA guidance, Appendix A.

seek to centralize IT functions, they must support statistical agencies' ability to control access to and use of their confidential data to ensure that the data are kept confidential and used only for statistical purposes.[9]

Although statistical agencies must be able to make pledges of confidentiality, it is not required that they do so for all of their collections. Statistical agencies may collect aggregated data from state and local governments that are already publicly available, and it would not serve the public good for the agency to then keep such data confidential.[10] Because it is expected that statistical agencies will collect data solely for statistical purposes with pledges of confidentiality, they must be very clear when any data they are collecting will have nonstatistical uses.[11]

The authority of a statistical agency to release statistical information (including press releases) without prior clearance for the statistical content by department policy officials is essential, so that there is no opportunity for or perception of political manipulation of any of the reports.[12] Statistical agencies are required to adhere to predetermined schedules for the public release of key economic indicators and to take steps to ensure that no person outside the agency has prior access except under carefully specified conditions.[13] Agencies are also required to develop and publish schedules for the release of other important social and economic indicators and to announce and explain any changes in schedules as far in advance as possible.[14]

Statistical agencies are encouraged to use press releases to expand the dissemination of data to the public. However, such press releases must "be produced and issued by the statistical agency and must provide a policy-neutral description of the data."[15] Any policy pronouncements must be issued separately by executive branch policy officials and not by the statistical agency, and "policy officials of the issuing department may review

---

[9]See 44 USC § 3563(b).

[10]The Census Bureau collects data from state and local governments in the Census of Governments without a pledge of confidentiality, but only uses the information for statistical purposes. The National Center for Education Statistics (NCES) collects some data on public schools that it makes publicly available and does not promise to keep the data confidential.

[11]See Confidential Information Protection and Statistical Efficiency Act (CIPSEA) guidance (Confidential Information and Statistical Efficiency Act of 2018, 2019), Appendix A.

[12]The Energy Information Administration had its independence authorized in this regard in Section 205 of the Department of Energy Organization Act of 1977; 42 USC § 7135(d): "The Administrator [of EIA] shall not be required to obtain the approval of any other officer or employee of the Department in connection with the collection or analysis of any information; nor shall the Administrator be required, prior to publication, to obtain the approval of any other officer or employee of the United States with respect to the substance of any statistical or forecasting technical reports which he has prepared in accordance with law."

[13]See *Statistical Policy Directive No. 3* (OMB, 1985) in Appendix A.

[14]See *Statistical Policy Directive No. 4* (OMB, 2008) in Appendix A.

[15]See *Statistical Policy Directive No. 4* (OMB, 2008) in Appendix A.

the draft statistical press release [solely] to ensure that it does not include policy pronouncements."[16]

Statistical agencies should also have dissemination policies that foster regular, frequent release of major findings from the agency's programs to the public through the traditional media, the Internet, and other means. They should also provide access to the underlying data, using appropriate safeguards to protect confidentiality (see Practice 9) to permit their results to be replicated. In these ways, an agency can guard against the perception of political and other undue external influence that might bias its operations.

The head of the statistical agency or unit should be able to meet with congressional staff and members to explain the agency's statistics and programs. Although department representatives may also attend these meetings, the department should fully support the statistical agency in this regard. The head of the statistical agency or unit should also be able to prepare a budget request specific to their agency and meet with OMB during the annual budget development process.[17] Similarly, it is essential that statistical agency leadership and staff be able to interact directly with their users and stakeholders. While the department may benefit from hearing the needs and concerns of these groups and individuals, the statistical agency should have the autonomy to arrange these meetings.

Finally, statistical agencies should be vigilant to threats to their independence, but they should also seek to educate officials in their ecosystem proactively about the appropriate roles and responsibilities of a statistical agency. Statistical agencies, not their parent agencies, should be given clear credit or recognition for their data/reports. Senior leaders of an agency should cite relevant laws, regulations, and these widely accepted principles and practices for federal statistical agencies as precedent and as necessary for the mission of the agency. Undermining the authorities described in this practice undermines the mission of the agency itself, so if serious threats are made to a statistical agency's independence and references to the relevant laws, regulations, principles, and practices are not heeded, senior leaders should turn to the secretary of the department, the Chief Statistician of the U.S. at OMB, Congressional oversight committees, stakeholders, professional associations, and users to come to the agency's defense. Such outreach should not be undertaken lightly but should not be avoided if the fundamental mission of the agency is at stake.

---

[16] See *Statistical Policy Directive No. 4* (OMB, 2008) in Appendix A.

[17] The Office of the Chief Statistician of the U.S. is responsible for assessing federal statistical agency budgets under the Paperwork Reduction Act (Paperwork Reduction Act, 1995). Language at § 1321.4(g)(2) in the final Trust Regulation (OMB, 2024b) requires federal statistical agencies and units to be given the opportunity to participate in program and staffing budget preparation and their engagement during the annual budget review process.

PRACTICE 3: COMMITMENT TO QUALITY AND PROFESSIONAL STANDARDS OF PRACTICE

A federal statistical agency's commitment to quality and professional standards is the foundation of its credibility; such a commitment provides a strong defense against baseless attacks on the quality of statistical products and other forms of intentional disinformation. Such commitment should be deeply embedded in the agency's culture and reflected through:

1. Adhering to and implementing OMB standards and guidelines;
2. Publishing and implementing agency quality standards;
3. Maintaining quality assurance programs and innovating to improve data quality and the processes of compiling, editing, documenting, analyzing, and disseminating data;
4. Evaluating the quality of the agency's data (see Practice 6);
5. Communicating clearly what is known about the validity and accuracy of the agency's data and the resulting measures of quality (both uncertainty and bias; see Practice 10);
6. Documenting and updating concepts, definitions, and data collection methods and possible sources of error in data releases to the public (see Practice 10); and
7. Developing and maintaining relationships with appropriate professional organizations in statistics and relevant subject-matter areas (see Practice 5).

An effective statistical agency devotes resources to developing, implementing, and updating standards for data quality and professional practice. Although a long-standing culture of data quality contributes to professional practice, an agency should document standards through an explicit process. Having explicit standards that are regularly reviewed and updated facilitates the training of new in-house staff and contractors' staffs. The reviews should include a careful consideration of quality frameworks used by other national statistical organizations as well as international organizations (see Appendix C).

To ensure the quality of its data collection programs and data releases, an effective statistical agency combines formal quality assurance programs with mechanisms and processes for obtaining both inside and outside reviews (see Practice 6). Formal quality assurance programs include well-developed methods for detecting outliers and other errors in raw data, methods for identifying errors from editing and other data processing steps, and, increasingly, reviews of processes followed by holders of administrative and private-sector input data. Reviews help ensure data quality by addressing various aspects of an agency's operations, including the soundness of the data collection and estimation methods and the completeness of the documentation of the methods used, metadata, and the error properties of the data. For individual reports, formal processes are needed that incorporate review by agency technical experts and, as appropriate, by technical experts in other agencies and outside the government.[18]

An effective statistical agency keeps up to date on developments that may be relevant to its program—for example, modern methods for combining data from more than one source, use of artificial intelligence methods, estimating error in the resulting statistics, and new technologies for data collection, processing, and dissemination.

Statistical agencies should be alert to social and economic changes that may call for innovations in the concepts or methods they use (NASEM, 2017e, 2019a, 2020b). The need for change often conflicts with the need for comparability with past data series. Agencies have the responsibility to manage this conflict by initiating more relevant series or revising existing series to improve quality, while providing information to compare old and new series.

The best resource for ensuring high-quality data is a strong professional staff, which includes experts in the subject-matter fields covered by the agency's programs, experts in statistical methods and techniques, and experts in data collection, computing and information science, and other operations (see Practice 4). A major function of an agency's leadership is to strike a balance among these staff and to promote collaboration, with each group of experts contributing to the work of the others. An effective statistical agency encourages its professional staff's membership and participation in relevant professional associations to refresh their skills and knowledge and to develop networks of experts from other statistical agencies, academia, and the private sector (see Practice 4).

An effective statistical agency also has policies and practices to instill the highest possible commitment to professional ethics among its staff. Because knowledge of codes of ethics from professional associations can reinforce this commitment in the agency culture (Hogan & Steffey, 2014), an effective agency ensures that its staff members are aware of and have access to such statements of professional ethical practice as those of the American Association for Public Opinion Research (2021),[19] the American Statistical Association (2018 and 2024),[20] the American Economic Association (2018),[21] and the International Statistical Institute (2023),[22] as well as the agency's own policies and practices regarding such matters as the

---
[18] See (OMB, 2005).
[19] https://aapor.org/standards-and-ethics/
[20] https://www.amstat.org/asa/files/pdfs/EthicalGuidelines.pdf and https://www.amstat.org/policy-and-advocacy/asa-board-statements
[21] https://www.aeaweb.org/ethics
[22] https://isi-web.org/declaration-professional-ethics

protection of confidentiality, respect for privacy, and standards for data quality. This reinforcement of professional and agency ethical cultures is recognized by the federal Data Ethics Tenets (General Services Administration, 2020). An effective agency endeavors in other ways as well to ensure that its staff are fully cognizant of the ethics that must guide their actions in order for the agency to maintain its credibility as a source of objective, reliable information for use by all (Hartman et al., 2014).

PRACTICE 4: PROFESSIONAL ADVANCEMENT OF STAFF

The long-term credibility of a statistical agency depends on the agency's staff and the culture it builds and maintains for quality and professionalism. Thus, a statistical agency should recruit and support highly qualified and dedicated staff for all aspects of its operations, including subject-matter experts in fields relevant to its mission (e.g., demographers, economists), statistical methodologists who specialize in data collection and analysis, computer and data scientists, and other skilled staff such as budget analysts, procurement specialists, and human resource specialists. Statistical agency staff should be recruited and promoted based solely on their professional qualifications and performance, and these personnel decisions should be made solely by agency career staff without external interference (see Practice 2). Statistical agencies also should consider the pipeline of future professionals to ensure the long-term viability of statistical programs and products. This includes finding ways to facilitate the training and development of future generations of professionals needed to design and manage statistical systems of the future.

To manage its staff effectively, an agency should provide them with opportunities for work on challenging projects in addition to more routine, production-oriented assignments. An agency's personnel policies, supported with sufficient resources, should enable staff to extend their technical capabilities through appropriate professional and developmental activities (see below). These activities enhance the knowledge and skills of the staff members and pay dividends to the agency, helping it to stay on top of new developments.

The personnel policies of an effective federal statistical agency should encourage the development and retention of a strong professional staff who are committed to the highest standards of quality work for their agency and in collaboration with other agencies. Key elements of such policies include the following:

1. Providing staff with continuing technical education and training, appropriate to the needs of their positions. Technical education may come from in-house training programs and opportunities for

external education and training at universities or through professional societies. Supervisory and leadership training from the U.S. Office of Personnel Management or other institutions should also be encouraged for managers and emerging leaders;
2. Structuring position responsibilities to ensure that staff have the opportunity to participate, in ways appropriate to their experience and expertise, in research and development activities to improve the quality of data and cost-effectiveness of agency operations;
3. Encouraging and recognizing professional activities, such as publishing in refereed journals and presenting at conferences. The latter should include technical work in progress, with appropriate disclaimers;
4. Supporting participation in relevant statistical and other scientific associations and committees, including leadership positions, to promote interactions with researchers and methodologists in other organizations that can advance the state of the art. Such participation is also a mechanism for disseminating information about an agency's programs and helps ensure a culture of scientific integrity within the agency;[23]
5. Fostering interaction with other professionals inside and outside the agency through a variety of mechanisms, for example participation in technical advisory committee meetings, interaction with contract researchers and research consultants on substantive matters, interaction with visiting fellows and staff detailed from other agencies, developmental assignments with other relevant statistical, policy, or research organizations, and rotational assignments within the agency;
6. Exploring opportunities to engage experts for short duration projects at federal statistical agencies to share and apply the latest statistical and data science techniques through existing authorities, such as through the Intergovernmental Personnel Act (Cui et al., 2024, forthcoming; Ho & O'Connell, 2024; Intergovernmental Personnel Act of 1970, 1970; Temporary Assignments Under the Intergovernmental Personnel Act, 2024) and Excepted Service (Excepted Service, 1978);
7. Supporting participation in cross-agency collaboration efforts, such as the Federal Committee on Statistical Methodology and its subcommittees. Such participation not only benefits the professional staff of an agency, but also contributes to improving the work of the statistical system as a whole (see Practice 7);

---

[23] See Office of Science and Technology Policy Memorandum on Scientific Integrity (Office of Science and Technology Policy, 2023b) in Appendix A.

8. Rewarding accomplishment by appropriate recognition and by affording opportunities for further professional development. The prestige and credibility of a statistical agency is enhanced by the professional visibility of its staff, which may include establishing high-level nonmanagement positions for highly qualified technical experts; and
9. Seeking opportunities to reinforce the commitment of its staff to ethical standards of practice.

Implementing these policies requires sufficient funding, time off, and institutional respect for professional education and development.

An effective statistical agency carefully considers the costs and benefits—both monetary and nonmonetary—of using contractor organizations, not only to collect data but also to supplement in-house staff in other areas, such as carrying out methodological research. Outsourcing can have benefits, such as providing expertise in areas in which the agency is unlikely to be able to attract highly qualified in-house staff (e.g., some information technology functions), enabling an agency to handle an increase in its workload that is expected to be temporary or that requires specialized skills, and allowing an agency to learn from best industry practices. However, over time excessive outsourcing can also have unintended costs, including a transformation of agency staff from being primarily technical experts in their fields to serving primarily as contract managers, with an associated loss of in-house knowledge. (See, in particular, recommendations 5.1 and 5.2 in NASEM, 2022a.)

An effective statistical agency maintains and develops a sufficient corps of in-house staff, including mathematical statisticians, survey researchers, subject-matter specialists, data and computational scientists, and information technology experts, who are qualified to analyze the agency's data and to plan, design, carry out, and evaluate its core operations, so that the agency maintains the integrity of its data and its credibility in planning and fulfilling its mission. Agencies also need staff with specialized skills to create visualizations, metadata, and application programming interfaces for data dissemination (see Practice 9). At the same time, statistical agencies should maintain and develop staff with the expertise necessary for effective technical and administrative oversight of contractors. Given the increasing use of alternative data sources, agencies should not only encourage training in programming and software engineering to build up their staff's skills in data science, but also encourage their subject-matter experts to become fully knowledgeable about the origin, content, and quality of various relevant data sources.

Having sufficient in-house staff with the required types of expertise is as critical as having adequate budget resources for enabling a statistical agency to carry out its mission. Statistical agencies are constrained by federal personnel policies that can affect whom they are permitted to hire (e.g., U.S. citizens) and by federal pay scales. However, some statistical agencies have been needlessly constrained in the number of agency staff they can employ regardless of their budgetary resources, resulting in too few staff to adequately handle the work needed to maintain existing programs and oversee contractors (see American Statistical Association, 2024). As part of their fundamental responsibilities to support statistical agencies, departments housing statistical agencies should work with and support them in being able to hire a sufficient number of staff with the right expertise to carry out their missions.

PRACTICE 5: AN ACTIVE RESEARCH PROGRAM

Statistical agencies need active research programs that are closely tied to their mission of producing relevant and high-quality statistics. Research is not an "optional" or "extra" activity that can be deferred whenever resources are tight. It produces the innovation that refreshes relevance. The underfunding of statistical agencies' research has threatened the data infrastructure that provides vital information needed by governments, businesses, organizations, and individuals.[24]

To maintain relevance for public and policy purposes, federal statistical agencies must identify emerging needs and look for ways to develop new information sources. To improve the quality and timeliness of their data products, they must keep abreast of methodological and technological advances and be prepared to implement new procedures in a timely manner (see Practice 3). They must also continually seek ways to make their operations more efficient (see Practice 6).

An effective statistical agency's research program includes research on the substantive issues for which the agency's data are compiled as well as methodological research to improve statistical methods and operational procedures. Questions related to the use of administrative records and alternative data sources to enhance or potentially replace some of the information currently obtained through surveys have been a focus of research for statistical agencies for decades, which is only growing in importance. These questions include how closely statistics from these administrative and alternative data sources correspond to existing measured concepts, what additional information they may offer, and the appropriate methodologies for evaluating quality and integrating data sources.

---

[24]https://www.linkedin.com/pulse/federal-statistical-agencies-struggle-maintain-vitalrole-citro/?trackingId=hWmaUxpC4ao5VxtMmWioyg%3D%3D

Substantive Research and Analysis

A statistical agency should include staff with responsibility for conducting objective substantive analyses of the data that the agency compiles, such as analyses that assess trends over time or compare population groups. Substantive analyses provided by an agency should be kept relevant to policy by addressing topics of public interest and concern; however, such analyses should not espouse policy positions or be designed to reflect any particular policy agenda (Martin, 1981; Norwood, 1975; Triplett, 1991). The existence and output of an analytical staff can contribute not only to the knowledge base in the applicable subject areas, but also to the credibility, relevance, accuracy, timeliness, and cost-effectiveness of the agency's data collection programs. Benefits that a strong subject-matter staff bring to a statistical agency include the following:

1. Agency analysts understand the need for the data from a statistical program and how the data will be used, and they can communicate more effectively with data users (see Practice 9);
2. Agency analysts typically have access to the complete microdata and so are better able than outside analysts to understand and describe the limitations of the data for analytic purposes and to identify errors or shortcomings in the data that can lead to subsequent improvements (see Practice 10); and
3. Substantive research maintains the relevance of an agency's data program, suggesting changes in priorities, concepts, and needs for new data or discontinuance of outmoded or little-used series.

An agency's subject-matter analysts should be encouraged and have ample opportunity to build networks with analysts in other agencies, academia, the private sector, other countries, and relevant international organizations and to present their work at relevant conferences and in working papers and refereed journal articles (see Practice 4).

Research on Methodology and Operations

Statistical agencies should be innovative in the methods they use for data collection, processing, estimation, analysis, and dissemination, with the goals of improving data accuracy, timeliness, and operational efficiency and of reducing respondent burden. Careful evaluation of new methods is required to assess their benefits and costs in comparison with current methods and to determine effective implementation strategies, including the development of methods for bridging time series before and after a change in procedures.

Research on methodology and operations must be ongoing and geared to both current and future needs. Some current research topics are listed below.

1. Developing methods for producing rapid statistics to respond to high-priority situations or emergencies, such as the COVID-19 pandemic;[25]
2. Evaluating administrative records for use to replace or enhance existing surveys;
3. Investigating the use of artificial intelligence and related methods to improve estimation or processing;
4. Assessing uncertainty when combining data from a variety of data sources;
5. Examining administrative records and other data sources as a means to provide provisional subnational estimates;
6. Improving the accuracy of survey estimates in the presence of nonresponse;
7. Using adaptive designs for maintaining and improving the quality and the cost-effectiveness of surveys;
8. Understanding and minimizing mode effects on data quality; and
9. Developing and evaluating new methods of confidentiality protection.[26]

Surveys will likely remain an important component of federal statistical agencies' portfolios because (a) some information is best (or only) obtained by asking questions; and (b) surveys can collect information on many characteristics at the same time, thereby permitting rich multivariate analysis. But declining survey response rates are making it increasingly difficult to maintain high data quality while controlling data collection costs (NASEM, 2017a; NRC, 2013b). Many of the large federal surveys are designed to produce annual nationwide estimates and do not produce the rapid and granular estimates needed by some data users. It is thus essential to consider how administrative records and alternative data sources can bolster the completeness, quality, and utility of statistical estimates while containing costs and reducing respondent burden (OMB, 2016f).

---

[25]For example, see National Center for Health Statistics (NCHS) guidance in data collection and methodology for COVID-19 data at https://www.cdc.gov/nchs/covid19/index.htm; the NCES School Pulse Panel data collection and methodology described at https://nces.ed.gov/surveys/spp/ and *Statistical Agency Changes in Response to the COVID-19 Pandemic* at https://www.statspolicy.gov/assets/docs/ICSP-COVID-19-Report_011521.pdf

[26]The National Secure Data Service Demonstration Project at the National Center for Science and Engineering Statistics (NCSES) will be informed by several studies sponsored through the America's DataHub Consortium, to evaluate privacy enhancing technologies and their application to statistical products. See Practice 9.

Expanding the Statistical Use of Administrative Records

Administrative records include records of federal, state, and local government agencies that are used to administer a government program. Examples include U.S. Social Security Administration records of payroll taxes collected from workers and benefits paid out to beneficiaries; state agency records provided by applicants for assistance programs and payments to applicants deemed eligible; and local government property tax records. Administrative records have been used for statistical purposes for many years to generate up-to-date population estimates by age, gender, race, and ethnicity. In turn, these estimates are used to adjust population survey weights for coverage errors and for many other purposes (NRC, 2004d, 2007a).

Some of the many examples of statistical agencies' use of administrative data include the Census Bureau using tax records for the economic censuses for small and non-employer businesses,[27] the NCHS National Vital Statistics System drawing upon birth and death records from the states,[28] and the NCES' National Postsecondary Student Aid Study, drawing upon federal and institutional administrative data to analyze student financial aid.[29] Research is being conducted to assess whether tax records can replace income items in the American Community Survey (NASEM, 2019b). Administrative records are also frequently used with survey data to produce model-based estimates with improved accuracy for small geographic areas or population groups (NASEM, 2019b, 2023c; NRC, 2000a,b; Young, 2019; Young & Chen, 2022).

There are many other potential statistical uses for administrative records from program agencies, and expanding the use of these records could improve the cost-effectiveness and quality of some statistical programs. Potential uses include substituting administrative records for specific survey questions and adding richness to a combined dataset by appending administrative records variables to matched survey records (Commission on Evidence-Based Policymaking, 2017; NASEM, 2018a, 2019c, 2023b,c; NRC, 1997b, 2009a, 2012a). Administrative records from multiple federal agencies are also being used in the decennial census to verify vacant units and, when good information exists, to fill in data if an initial nonresponse follow-up visit is not successful in locating a respondent.[30]

---

[27]Non-employer businesses include just the sole proprietor with no other employees.
[28]See https://www.cdc.gov/nchs/nvss/index.htm.
[29]See https://nces.ed.gov/surveys/npsas/index.asp.
[30]See https://www2.census.gov/programs-surveys/decennial/2020/program-management/planning-docs/administrative-data-use-2020-census.pdf.

Evaluating and Using Alternative Data Sources

This data-rich age has a multitude of data sources beyond administrative records, including data gleaned or "scraped" from internet websites (e.g., price quotes, social media postings), data extracted from sensors (e.g., from traffic cameras), and data obtained from the private sector (e.g., credit card transactions, scanner data on retail purchases). Often, these sources generate large volumes of data that require computationally intensive techniques for extracting useful information for statistics (NASEM, 2017a,c, 2023c; NRC, 2008b). However, to make use of most nontraditional data sources, it is necessary for statistical agencies to first evaluate the accuracy and error properties of the data, and then to compare error magnitude and impact between alternative and traditional data collection methods.

In an era when data users expect timeliness and when budgets are constrained, researchers in statistical agencies should explore how nontraditional data sources can contribute to their programs (NASEM, 2017a,c, 2023b,c). Procedures could include (a) augmenting information obtained from traditional sources; (b) replacing information elements previously obtained from traditional sources; (c) providing preliminary estimates that are later benchmarked with traditional sources; and (d) analyzing information streams to identify needed changes (e.g., in types of jobs, education majors) in statistical classifications and survey questions (NASEM, 2017a,c, 2023b,c). A major challenge for statistical agencies has been the difficulty of identifying, locating, and accessing alternative data sources that could be useful for their programs. As the Evidence Act (Foundations for Evidence-Based Policymaking Act of 2018, 2019) is implemented, the data inventories and practices of the program agencies should continue to make these resources more transparent and make processes for obtaining these datasets for statistical purposes more streamlined (also see Practices 8 and 9).

In considering their strategies, statistical agencies should adopt broad quality frameworks that capture user needs, including aspects such as relevance, accuracy, timeliness, comparability (over time and with other data sources), transparency, accessibility, privacy, protection from outside manipulation, and interpretability. They should examine the tradeoffs between different quality aspects, such as trading precision for timeliness and granularity (see NASEM, 2017c,e, 2023c, 2024c; Appendix C). An agency's own research staff can assist in examining these tradeoffs, and the Federal Committee on Statistical Methodology (Federal Committee on Statistical Methodology, 2020; Prell et al., 2019) also has been pursuing work in this area to assist agencies.

Value of an Active Research Program

Supporting federal agencies' in-house research staffs is critical given the challenges and opportunities posed by the increasing availability of alternative data sources. Many current practices in statistical agencies were developed through research they conducted or obtained from other agencies. Federal statistical agencies, frequently in partnership with academic researchers, pioneered the use of statistical probability sampling, the national economic accounts, input-output models, and other analytic methods. The Census Bureau pioneered the use of computers for processing the census. Several statistical agencies use academic principles of cognitive psychology—a research strand dating back to the early 1980s (see NRC, 1984)—to improve the design of questionnaires, the clarity of data presentation, and the ease of use of electronic data collection and dissemination tools. History has repeatedly shown that research conducted within federal statistical agencies on subject areas, methods, and operations can lead to large productivity gains in statistical activities at relatively low cost (American Statistical Association, 2024; Citro, 2016; NRC, 2010b).

An effective statistical agency also actively partners with the academic community and private sector for methodological research. It seeks out academic and industry expertise for improving data collection, processing, and dissemination operations. For example, a statistical agency can learn techniques and best practices for improving software development processes from computer scientists (NRC, 2003a, 2004c). An effective agency also learns from and contributes to methodological research of statistical agencies in other countries and relevant international organizations (see Practice 7). Thus, it is important for agency staff to seek to publish their work in the leading peer-reviewed journals, and to post white papers and reports on agency websites, both of which enable broader dissemination as well as adding credibility to the changes the agency makes.

Preparing for the future requires that agencies periodically assess the scope of existing data series, alter data series as required, and innovate to improve their programs. Because of the decentralized nature of the federal statistical system, innovation often requires and benefits from cross-agency collaboration (see Practice 7) and a willingness to implement different kinds of data collection efforts to answer different needs, while being mindful of the need for historical trend data and comparability across different levels of geography. As described under Practice 7, a significant role of the Office of the Chief Statistician of the United States is to support cross-agency collaboration.

PRACTICE 6: STRONG INTERNAL AND EXTERNAL EVALUATION PROCESSES FOR AN AGENCY'S STATISTICAL PROGRAMS

Statistical agencies should have processes in place to support regular evaluations of their major statistical programs and their overall portfolio of programs. Reviews of major data collection programs and their components should consider how to produce relevant, accurate, and timely data in the most cost-effective manner possible. Reviews of an agency's portfolio should consider ways to reduce duplication, fill gaps, and adjust priorities so that the overall portfolio is as relevant as possible to the information needs of policymakers and the public (NASEM, 2018c, 2020b; NRC, 2009b). Such evaluations should include internal reviews by staff and external reviews by independent groups.

Agencies should seek administrative and outside reviews not only of specific statistical programs but also of program priorities and quality practices across their entire portfolio. They should also consider ways to improve program cost-effectiveness by combining data from multiple sources, particularly because fewer people and organizations are responding to surveys than in the past. It is increasingly urgent to determine whether there are alternative data sources to surveys that offer similar or better quality. (See NASEM, 2017a,b,c, 2018b, 2022a, 2023c; and Practice 5).

Statistical agencies that fully follow practices related to an active research program (Practice 5), openness (Practice 10), dissemination of statistical data products (Practice 9), and commitment to quality and professional standards (Practice 3) will likely be in a good position to make continuous assessments of and improvements in the relevance, quality, and efficiency of their data collection systems. Yet even the best-functioning agencies will benefit from an explicit program of internal and independent external evaluations to formalize success criteria, evaluate performance, and obtain fresh perspectives.

Evaluating Quality, Relevance, Efficiency

Evaluation of data quality for any kind of data collection program begins with regular monitoring of quality indicators that are readily available to users. Agencies should use broad quality frameworks (see Practice 3 and Appendix C) and assess the costs and benefits of using alternative data sources (see Practice 5 and NASEM, 2017a,c). These evaluations should be undertaken periodically and the results made public (see Practice 10 and NRC, 2007a).

When it is disruptive to implement improvements on a continuing basis, a common practice is to bundle changes to implement several at the same time. For example, classifications such as the North American Industry

Classification System (NAICS) are updated every 5 years, and agencies may implement other changes at the same time as this. Agencies should ensure that the intervals between innovations do not become so long that data collection programs deteriorate in quality, relevance, and efficiency. Regular, well-designed program evaluations, with adequate budget support, are key to ensuring that data collection programs do not deteriorate. Having a set schedule for research and development efforts will enable data collection managers to ensure that the quality and usefulness of their data are maintained and help prevent locking in less optimal procedures.

As part of ongoing evaluation, the relevance of an agency's data collection programs and products needs to be continually assessed. The question of relevance is whether the agency is "doing the right thing," in contrast to whether the agency is "doing things right." Relevance should be assessed not only for particular programs or closely related sets of programs, but also for an agency's complete portfolio in order to make the best choices among program priorities given the available resources (see Practice 1).

Engaging and consulting with stakeholders—through such means as regular meetings, workshops, conferences, and other activities—is important to ensuring relevance (see Practice 9). Including other federal statistical colleagues in this communication, both as users and as collaborators, can be valuable (see Practice 7).

Finally, statistical agencies should review their statistical programs for efficiency and cost-effectiveness.[31] Federal statistics as a public good represent a legitimate draw on public resources, and statistical agencies in turn are properly called on to analyze the costs of their programs on a continuing basis to ensure the best return possible on tax dollars. For this purpose, statistical agencies should develop complete, informative models for evaluating the costs of current procedures and of possible alternatives and follow best practice in the design of statistical production processes.[32]

Types of Reviews

Regular statistical program reviews should include a mixture of internal and external evaluation. Agency staff should set goals and timetables for internal evaluations that involve informed staff outside the program under

---

[31]"Efficiency" is generally defined as an ability to avoid waste (of materials, energy, money, time) in producing a specified output. "Cost-effectiveness" connotes a broader, comparative look at inputs and outputs to assess the most advantageous combination. ("Cost-benefit" analysis attempts to add monetary values to outputs.) In the context of federal statistical programs, cost-effectiveness analysis would assess the costs of conducting a program for different combinations of desired characteristics, such as improved accuracy or timeliness and reduced burden on respondents.

[32]See Generic Statistical Business Process Model in Appendix C.

review. Independent external evaluations should also be conducted on a regular basis. The frequency of these external evaluations should depend on the importance of the data, how quickly the phenomena being measured change, and how quickly respondent behavior and data collection technology may adversely affect a program change.

External reviews can take many forms. They may include recommendations from advisory committees that meet at regular intervals (typically, every 6 months). However, advisory committees should never be the sole source of outside review because the members of such committees rarely have the opportunity to become deeply familiar with agency programs. External reviews can also take the form of a special committee or panel established by a relevant professional association, such as the American Statistical Association, or by some other recognized group, such as the National Institute of Statistical Sciences or the Committee on National Statistics (also see NRC, 2009b).

Sunsetting Statistical Products or Programs

Although it can be difficult to stop producing something, all statistical agencies must be able to do so. While most reviews serve to improve ongoing programs, others may inform the decision to discontinue a particular product or program. The situations that call for sunsetting include an innovation or experiment that fails by some criterion, a budget reduction, a decline in relevance to users, a higher priority need elsewhere, or replacement by an alternative source (NASEM, 2022d). By informing such decisions, effective evaluations within agencies support continuous learning and program improvement in the federal statistical system and promote trust in the agencies.

An agency that never ends programs or products will eventually cease innovating and need to reduce the quality of all its programs. For example, in the face of a budget cut, discontinuing a program or product is preferable to across-the-board cuts in all programs, which would reduce the accuracy and usefulness of both the more relevant and less relevant data series (NASEM, 2022a). More generally, prudent use of taxpayer dollars requires that agencies be ready and able to sunset programs whenever resources could be more productively used elsewhere.

On considering a decision to terminate, a product or program communication is critical. The agency should be transparent to stakeholders, as far in advance as possible, about the nature of the change under consideration, the reasons for it, and alternative sources of information.

PRACTICE 7: COORDINATION AND COLLABORATION WITH OTHER AGENCIES

Statistical agencies should not be islands unto themselves. They need to engage and collaborate not only with their stakeholders but also with other statistical agencies and units in the federal government, in state governments, and internationally. The U.S. federal statistical system consists of many agencies in different departments, each with its own mission and subject-matter focus (see Appendix B). Yet these agencies have a common interest in serving the public need for credible, relevant, accurate, and timely information gathered as efficiently and fairly as possible. Moreover, needed information may often span the mission areas of more than one statistical agency: for example, both the Bureau of Labor Statistics (BLS) and NCES have programs that relate to education and employment outcomes. Consequently, statistical agencies should not and do not conduct their activities in isolation.

An effective statistical agency actively seeks opportunities to conduct research and carry out other activities in collaboration with other agencies to enhance the value of its own information and that of the system as a whole. Such collaboration is essential not only for smaller statistical agencies with limited staff and resources but, equally, for larger agencies so that they do not overlook useful innovations outside their own agency. When possible and appropriate, federal statistical agencies should collaborate not only with each other but also with policy, research, and program agencies in their departments, with state and local statistical agencies, and with foreign and international statistical agencies.

Such collaborations can serve many purposes, including standardization of concepts, measures, and classifications (see, e.g., NRC, 2004a,e; Appendix A); augmentation of available information for cross-national and subnational comparisons (NRC, 2000a,b); identification of useful new data sources and data products; and improvements in many aspects of statistical programs.

In their new roles as the chief Statistical Officials for their departments, heads of the recognized statistical agencies and units should proactively collaborate with their departments' Chief Data Officers, Chief Information Officers, Chief Artificial Intelligence Officers, and Evaluation Officers in overseeing departmentwide data governance and use of data for evidence building. Departments should consult the Statistical Officials during decision making, as statistical agencies add value by contributing expertise in areas such as data standards, privacy protection, rigorous methods for developing credible data that are fit for purpose, and appropriate interpretation of evidence and statistics (NASEM, 2022a). In addition, statistical agency heads should help improve statistical practices not only within their departments, but also across government by providing advice and mentoring, as appropriate, in particular to the Statistical Officials who are not heads of a recognized statistical agency or unit. This type of collaboration can help foster the understanding and integration of principles and practices for statistical activities that will be more widely adopted across the federal government as agencies continue to implement the Evidence Act (Foundations for Evidence-Based Policymaking Act of 2018, 2019; NASEM, 2022a), the Trust Regulation (OMB, 2024b), and the Information Quality Act of 2000 (Information Quality Act, 2000).

Coordinating Role of the Office of Management and Budget

The responsibility for coordinating statistical work in the federal government is specifically assigned to the U.S. Chief Statistician, who leads the Statistical and Science Policy (SSP)[33] Office in the Office of Information and Regulatory Affairs in OMB (see Appendix A). The U.S. Chief Statistician chairs the Interagency Council on Statistical Policy (ICSP), which consists of the heads of the recognized statistical agencies and units, and other agency Statistical Officials, to coordinate federal statistical programs and activities across the federal government (see Appendix B).

A primary responsibility of the Office of the U.S. Chief Statistician (also known as the SSP of OMB) is to identify issues of common concern and create interagency committees for collaborative work, such as concepts of interest to more than one agency (e.g., classifications of sexual orientation, gender identity, and race/ethnicity), the development and periodic revision of standard classification systems (e.g., of industries, products, occupations, and metropolitan areas), and best practices for domains such as survey methods, statistical use of administrative records, and confidentiality protection. SSP may then take the recommendations of these committees and issue more formal guidance or directives for all agencies to follow. To facilitate this coordination, some experts have called for a systemwide strategic plan that could assist the Office of the U.S. Chief Statistician in communicating its priorities and assessing the capacity of the federal statistical system (American Statistical Association, 2024).

Forms of Interagency Collaboration

Interagency collaboration and coordination take many forms, some multilateral, some bilateral. Some collaborations are formally chartered by OMB or the ICSP to perform a specific task, while others result from

---

[33] Officially titled the Statistical and Science Policy Office; also known as the Office of the Chief Statistician of the U.S.

common interests and may continue for years as a means of sharing information. Some interagency collaborations have been active for decades. Since 1975, the Federal Committee on Statistical Methodology (Federal Committee on Statistical Methodology, 2024), chaired by SSP, has convened technical experts across the federal government to advise OMB and the ICSP on methodological and statistical issues that affect the quality of federal data. This committee also provides a forum for statisticians in different federal agencies to discuss issues affecting federal statistical programs and promotes collaborative research.

Other ongoing collaborations, such as the Federal Interagency Forum on Aging-Related Statistics (Federal Interagency Forum on Aging-Related Statistics, 2024) and the Federal Interagency Forum on Child and Family Statistics (Federal Interagency Forum on Child and Family Statistics, 2024), provide statistical information to the public in a broad area of interest. These forums produce regular products that draw data from a wide range of agencies to provide a broad description of their population of interest in publications and materials that are easily understood and used by a broad audience.

A common bilateral arrangement is an agreement of a program agency to provide administrative data to a statistical agency to use as a sampling frame, a source of classification information, a summary compilation to check (and possibly revise) preliminary sample results, and a source with which to improve imputations for survey nonresponse, reduce variability in estimates for small geographic areas, or substitute for survey questions. The Census Bureau, for example, uses Schedule C tax information from the Internal Revenue Service in place of surveys for millions of nonemployer businesses. Such practices improve statistical estimates, reduce costs, and eliminate duplicate requests for information from the same respondents.

In other arrangements, federal statistical agencies engage in cooperative data collection with state statistical agencies to let one collection system satisfy the needs of both. A number of such joint systems have been developed, notably by BLS, the National Agricultural Statistics Service, NCES, and NCHS.

Another example of a joint arrangement is one in which one statistical agency contracts with another to conduct a survey, compile special tabulations, or develop models. Such arrangements make use of the special skills of the supplying agency and facilitate the use of common concepts and methods. The Census Bureau conducts many surveys for other agencies; both NCHS and the National Agricultural Statistics Service receive funding from other agencies in their departments to support their survey work; and NCSES receives funding from agencies in other departments to support several of its surveys.

International Collaborations

In order to be most relevant and useful, many federal statistics must be internationally comparable. The U.S. Chief Statistician at OMB is responsible for coordinating U.S. participation in international statistical activities. Many other agencies' staffs participate in a wide variety of activities in collaboration with other national statistical offices, such as working groups sponsored by the United Nations Statistical Commission and the Organization for Economic Co-operation and Development. These activities include participating in the development of international standard classifications and systems; supporting educational activities that promote improved statistics in developing countries; and learning from and contributing to the work of established statistical agencies in other countries in such areas as survey methodology, record linkage, confidentiality protection techniques, and data quality standards.

There are a growing number of international frameworks and tools that describe the common activities of statistical organizations and facilitate the documentation and sharing of data and metadata. The Generic Statistical Business Process Model describes and defines the set of business processes needed to produce official statistics. It provides a standard framework and harmonized terminology to help statistical organizations modernize their statistical production processes, as well as to share methods and components (see Appendix C). There is also ongoing international work on using administrative and big data sources for federal statistics and on the quality frameworks for these data sources.

Challenges and Rewards for Collaboration

Collaborative activities, such as sharing and integrating data compiled by different statistical and program agencies, standardizing concepts and measures, reducing unneeded duplication, and working together on methodological challenges, invariably require effort to overcome differences in agency missions and operations. There are also potentially greater threats to confidentiality because the linking of data provides more information that can lead to indirect identification. Yet with constrained budgets and increasing demands for more relevant, accurate, and timely statistical information, the importance of proactive collaboration and coordination among statistical agencies cannot be overstated. To achieve the most effective integration of their work for the public good, agencies must be willing to take a long view, to strive to accommodate each other, and to act as partners in the development of statistical information for public use. The rewards of effective collaboration can be not only data that are more efficiently obtained, of higher quality, and more relevant to policy concerns, but also a stronger, more effective statistical system.

Statistical agencies that collect similar information should consider integrating their microdata records for specified statistical uses as another way to improve data quality, develop new kinds of information, and increase cost-effectiveness. One cost-effective approach is for a large survey to provide the sampling frame and additional content for a smaller, more specialized survey. The National Health Interview Survey run by NCHS of the Centers for Disease Control and Prevention serves this function for the Medical Expenditure Panel Survey of the Agency for Healthcare Research and Quality. Similarly, the American Community Survey serves this function for the National Survey of College Graduates, which the Census Bureau conducts for NCSES (NRC, 2008c). The Office of the Chief Statistician of the U.S., in its established role of reviewing federal agency information collection requests to evaluate public benefit given respondent burden, identifies opportunities for cross-agency frame and content sharing. The Trust Regulation calls for further engagement and consultation across federal statistical agencies to better identify and address sources of duplication, and thereby reduce respondent burden (OMB, 2024b). As the data sharing provisions of the Evidence Act are implemented through the regulatory process, the Office of the U.S. Chief Statistician could further identify and make progress on opportunities for sharing through convening an interagency workgroup.

Another key collaboration is with states. Many federal statistical agencies have relationships with states for data collection, but they would also like greater access to state administrative records. Obtaining such access often depends on having good relationships with states (Goerge, 2018; NASEM, 2017a, 2024c).

Unfortunately, there are sometimes legal or administrative barriers that prevent statistical agencies from collaborating on common activities. Both BLS and the Census Bureau maintain business establishment lists, but each of the lists derives from different sources (state employment security records for BLS and a variety of sources, including federal income tax records, for the Census Bureau). Research has demonstrated that synchronizing the lists would improve the accuracy of the information and the coverage of business establishments in the United States (NRC, 2006a, 2007b). However, business establishment lists cannot currently be synchronized between BLS and the Census Bureau, partly because the latter is prohibited by law (Internal Revenue Code of 1986, 1986) from sharing with BLS (or Bureau of Economic Analysis) any tax information on businesses or individuals that it may acquire from the Internal Revenue Service, even for statistical purposes. Previous National Academies reports have recommended that these barriers be removed, as they negatively impact data quality and efficiency (NASEM, 2017a; NRC, 2007b). The future publication of regulations implementing the Evidence Act and identification of these barriers should lead to legal and administrative changes that will encourage joint statistical activities.

PRACTICE 8: RESPECT FOR DATA SUBJECTS AND DATA HOLDERS AND PROTECTION OF THEIR DATA

Federal statistical agencies are able to produce useful statistical information because they can collect and acquire data from data subjects and data holders, including survey respondents, organizations that provide data files, government agencies that provide administrative records, third-party data aggregators, and others. A statistical agency's ability to fulfill its mission thus depends upon the relationships that the agency is able to build and maintain with data subjects and data holders. Effective statistical agencies demonstrate respect for their data subjects and data holders and protect their data to ensure that agencies can fulfill their missions.

To maintain a relationship of respect and trust with data subjects and data holders, a statistical agency should respect their privacy, minimize the reporting burden imposed on them, and respect their autonomy when they are asked to participate in a voluntary program to collect data. The statistical agency must also comply with all legal requirements to ensure that the data are used only for statistical purposes. To do this, a statistical agency must communicate its privacy and confidentiality protection procedures and policies,[34] as well as the societal benefits from collecting the data.

Respecting Privacy in Surveys

To promote trust and encourage accurate responses from survey respondents, it is important that statistical agencies respect their privacy by reducing, to the extent possible, the intrusiveness of questions they ask, and the time and effort required to respond. Agencies must also give respondents adequate information with which to decide if a survey is worthy of response—that is, so respondents can give their informed consent (see below). Thus, when individuals or organizations are asked to participate in a survey, they should be told whether it is mandatory or voluntary, how the data will be used, and what confidentiality protections apply to the data. They should also be informed of the likely duration of a survey response

---

[34]Informational privacy is "an individual's freedom from excessive intrusion in the quest for information, and an individual's ability to choose the extent or circumstances under which his or her beliefs, behaviors, opinions, and attitudes will be shared with or withheld from others," while confidentiality refers broadly to an obligation not to transmit information to an unauthorized party (NRC, 1993a, p. 22).

task, whether they will be asked to consult records, and whether the survey involves repeated responses over time (OMB, 2016d).

To reduce the burden of replying to surveys (NRC, 2013c, Ch. 4; OMB, 2016f), statistical agencies should write clear questions that fit respondents' common understanding, minimize the intrusiveness of questions, and explain why intrusive-seeming questions serve important purposes. Statistical agencies should also allow alternative modes of response when appropriate (e.g., Internet, smartphone) and use administrative records or other data sources, if sufficiently complete and accurate, to provide some or all of the needed information. In surveys of businesses or other organizations, agencies should seek to obtain information directly from the organization's records and so minimize the need for duplicate responses to multiple requests. As described under Practice 7, the Office of the U.S. Chief Statistician reviews federal agency information collection requests to determine public value given respondent burden. Through this established process, opportunities for sharing sampling frames and content across agencies are identified, with an eye on reducing duplication. To further identify and make progress on these opportunities, the Office of the U.S. Chief Statistician could usefully convene an interagency working group.

Protecting and Respecting the Autonomy of Human Research Participants

Collecting data from individuals for research purposes using federal funds falls under a series of regulations, principles, and best practices that the federal government has developed over a period of more than 50 years (NRC, 2003b, 2014). The pertinent regulations, which have been adopted by 20 departments and agencies,[35] are known as the "Common Rule" (Federal Policy for Protection of Human Subjects, 2018). The Common Rule regulations, most recently revised effective January 2019, require that researchers adequately protect the privacy of human participants and maintain the confidentiality of data collected from them, minimize the risks to participants from the data collection and analysis, select participants equitably with regard to the benefits and risks of the research, and seek the informed consent of individuals to participate (or not) in the research. Under the regulations, most federally funded research involving human participants must be reviewed by an independent institutional review board (IRB) to determine whether the design meets ethical requirements for protection.[36]

---

[35] https://www.hhs.gov/ohrp/sites/default/files/revised-common-rule-reg-text-unofficial-2018-requirements.pdf

[36] For information about the Common Rule and certification of IRBs by the Office for Human Research Protections in the U.S. Department of Health and Human Services, see http://www.hhs.gov/ohrp and Appendix A.

Not all federal statistical agencies' data collections are subject to IRB review. Nonetheless, agencies should strive to incorporate the spirit of the Common Rule in the design and operation of all activities that involve data collection from individual respondents. Statistical agencies should seek ways to inform potential respondents that will help them decide whether to participate, such as sending respondents an advance letter. Such information should include the planned uses of the data and their benefits to the public.

Even for mandatory data collections, such as the decennial Census of Population and Housing and the quinquennial Economic Census, a statistical agency should respect its respondents by giving them as much information as possible about the reasons for the collection and making it as easy as possible for them to respond (OMB, 2016d). The principles and practices of respect apply not only to individuals asked to participate in a survey, but also to representatives of organizations (e.g., businesses, state, and local governments) asked to participate in a survey and to custodians of existing data, such as administrative records, who are asked to share their data for statistical purposes.

Respecting the Holders and Subjects of Administrative and Other Data

Moving to a new paradigm of using multiple data sources for federal statistics, an agency must develop procedures that respect the constraints of data holder organizations. In working with federal agencies that hold useful administrative records, statistical agencies should plan to cooperate, communicate, and coordinate with them on a continuing basis, as urged in Hendriks (2012). A continuing relationship of mutual respect and trust enables a statistical agency to better understand the strengths and limitations of data held by a custodial agency. Mutual respect can help identify improvements in the data that are useful to both agencies.

An important consideration in using administrative records is whether informed consent of the data subjects (whether individuals or organizations) that provided their information to the data holder agency is required. In many cases the statistical use of administrative records may qualify under the "routine use" exception of the Privacy Act to provide evidence for the effective operation of the program. In some instances, it may be necessary to obtain new consent from the original data subjects.

Protecting the Confidentiality of Data Subjects' Information

When individuals and organizations provide information to statistical agencies, they advance the public good. These data subjects must be able to rely on the statistical agency's promise to protect their information, to use it only for statistical purposes, and to protect it from other uses.

A credible pledge of confidentiality for data subjects is considered essential to encourage high response rates and accuracy of responses from survey participants.[37] Moreover, if data subjects have been assured of confidentiality, disclosure of identifiable information would violate the principle of respect even if the information is not sensitive and would not result in any social, economic, legal, or other harm. For sensitive administrative data obtained from another government agency data holder, there must be a credible pledge of confidentiality in a properly formulated memorandum of understanding or other authorizing document.

CIPSEA (Confidential Information Protection and Statistical Efficiency Act, 2002) was originally enacted in 2002 and recodified as part of the Evidence Act (Foundations for Evidence-Based Policymaking Act of 2018, 2019; see Appendix A). This law protects the confidentiality of all federal data collected for statistical purposes under a confidentiality pledge, including but not limited to data collected by statistical agencies.[38] CIPSEA thus provides a common basis for the protection of all statistical data across agencies, which enables some data sharing and provides statistical agencies the ability to designate external researchers as their agents in order to allow them access to data for statistical purposes. The law contains penalties for employees and agents who knowingly disclose identifiable statistical information (up to 5 years in prison, up to $250,000 in fines, or both).

Both the perception and the reality of agencies' confidentiality protection may be affected by departmental initiatives to consolidate data processing and storage to bolster computer and network security in the federal government, improve the cost-effectiveness of information technology development and maintenance, and protect against cyberattacks. An effective statistical agency will work with its department on approaches to computer security. As part of their responsibilities to support federal statistical agencies, departments should ensure that statistical agencies are able to control their data and information systems so that the data are only used for statistical purposes and are kept confidential (see Practice 2).

---

[37] See (Hillygus et al., 2006; NRC, 1979, 2004d, 2013b).

[38] Section 508 of the USA PATRIOT Act of 2001 (P.L. 107-56) amended the NCES Act of 1994 to allow the U.S. attorney general (or an assistant attorney general) to apply to a court to obtain any "reports, records, and information (including individually identifiable information) in the possession" of NCES that are considered relevant to an authorized investigation or prosecution of domestic or international terrorism. Section 508 also removed penalties for NCES employees who furnish individual records under this section. This exclusion for NCES has not been invoked.

PRACTICE 9: DISSEMINATION OF STATISTICAL PRODUCTS THAT MEET USERS' NEEDS

An effective statistical agency produces and disseminates statistical products that meet the needs of its users. Users' needs for information evolve as new tools for using data become available. Once statistical information is made public, it will be used in numerous ways, including ways not originally envisaged, and by numerous types of users, ranging from government officials (whether federal, state, or local) to media, activists, data wholesalers, academic scholars, students, and data subjects. A statistical agency should continually strive to obtain input from data users on its programs, products, and dissemination tools and methods. Understanding data users' needs and how they use data products is critical for making an agency's data services as relevant, accurate, timely, and accessible as possible.

To return full value to the populace, statistical agencies also have an important role to play in helping users understand the strengths and limitations of the agencies' data, and, when asked, to contrast the usefulness of those data to user needs as compared to data from other sources being considered in the public domain. Citizens who might be misled by untested or biased alternative sources of data will benefit from understanding that they can count on the statistical agencies to produce high quality, objective, and trustworthy information. Furthermore, as the statistical agencies model good practices, they provide a natural contrast to less reliable statistical practices and products.

There is also increasing demand for statistics at greater levels of granularity, whether for disaggregated geographies or subgroups of the population relevant for policymaking. Statistical agencies must find ways to meet these increasing data demands while balancing the needs of data subjects and data holders (see Principle 3; Bowen & Snoke, 2023; NASEM, 2023b,c, 2024c).

Keeping abreast of the interests of current and potential new users requires continual attention to changes in the relevant policy issues and social and economic conditions in a statistical agency's domain, as well as changes in technologies for data access. Statistical agencies should work with professional associations, institutes, universities, and scholars to determine the current and emerging needs of research communities. They should also work with relevant professional associations and other organizations to determine the needs of business and industry as well as state and local governments. Statistical agencies also should proactively explore the needs of users through advisory committees.

Individual persons, households, businesses, institutions, organizations, and government entities have provided the underlying source data for an agency's statistics. Furthermore, the public has paid for the data collection, compilation, and processing. In return, the information created with such data should be accessible in ways that make it as useful as possible to the largest number of users—for decision making, program evaluation, scientific research, and public understanding (see also the Federal Data Strategy (OMB, 2021a) in Appendix A).

A statistical agency should strive for the widest possible dissemination of the statistical products it produces, consistent with its obligations to protect confidentiality. The products should be clearly identified, easy to find and use, and well documented. They should adhere to the FAIR (findable, accessible, interoperable, and reusable) principles,[39] which are becoming an international standard with adherence often required by research funding organizations and increasing adoption by national statistical institutes (Cabrera et al., 2020; Group of Eight, 2013). Dissemination should be timely, and information should be made readily available on an equal basis to all users. Agencies should have data curation policies and procedures in place so that data are preserved, fully documented, and accessible for statistical use in future years.[40]

Statistical agencies disseminate two broad classes of products: products that are publicly available, such as statistical releases, analytical reports, infographics, and public use microdata, and restricted access products, such as datasets containing confidential information, which are available in Federal Statistical Research Data Centers (FSRDCs) or through other restricted arrangements.[41] Both of these broad dissemination classes of data products actually come from the same underlying survey, administrative, and private data sources. This interdependency is important to understanding the risks when balancing public data products and expanded access to restricted data, and the effort thus required to manage access to both dissemination classes appropriately. Nonetheless, the process of gaining access to restricted data is often very time-consuming and costly for those who request access (see Conclusion 2–3 in NASEM (2023a). This undermines the principle of equitable access to data by a variety of users.

Increasingly, statistical agencies are moving away from this binary system of access models.[42] "Tiered access is an application of data minimization, a key privacy safeguard for evidence building [...] Data minimization means giving access to the least amount of data needed to complete an

---

[39] See https://www.go-fair.org/fair-principles/

[40] Data curation involves the management of data from collection and initial storage to archiving (or deletion should the data be deemed of no further use—e.g., a data file that represents an initial stage of processing). The purpose of data curation is to ensure that information can be reliably retrieved and understood by future users.

[41] https://www.census.gov/fsrdc

[42] https://nces.ed.gov/fcsm/dpt/content/4

approved project. Tiered access uses a variety of controls to minimize data needed for a given project and thereby reduce risk of disclosure of confidential data" (Commission on Evidence-Based Policymaking, 2017, p. 38). The Standard Application Process (SAP; OMB, 2022a) implemented as part of the Evidence Act is designed to simplify and speed up the application process for access to restricted data, but more needs to be done to facilitate such access (NASEM, 2023c).

Public Statistical Data Products

Dissemination of aggregate statistics may take the form of regularly updated time series, cross-tabulations of aggregated characteristics of respondents, analytical reports, interactive maps and charts, infographics, profiles, fact sheets, and press releases providing key findings. Such products should be readily accessible through an agency's website, supplemented by more detailed tabulations and data tools.[43]

Statistical agencies are using a variety of tools that make it easier for users to discover, query, retrieve, and use statistics on their own. These include redesigned websites, new platforms, interactive tools and applications, customized tables, new mapping capabilities, and application program interfaces.

For publicly available data products, a statistical agency's dissemination program should include the following elements:

1. An established publications policy that describes, for each statistical program, the types of reports and other data releases to be made available, the formats to be used, the audience to be served, and the frequency and timing of release;[44]
2. A variety of avenues for disseminating information about data availability and upcoming releases;
3. Multiple data products (suitably processed to protect confidentiality), so that information can be accessed by users with varying skills and needs. Useful data products may include not only understandable maps, graphs, indicators, tables, and interactive data tools and applications on statistical agency websites that meet the needs of different user groups, but also public-use microdata samples when practicable;

---

[43] Note the Trust Regulation requires each recognized statistical agency or unit to maintain a distinctive, outward-facing website with its own domain name and with adequate control over the website content and management to uphold the fundamental responsibilities. See § 1321.4(e).

[44] See *Statistical Policy Directive No. 3* (OMB, 1985) and 4 (OMB, 2008) in Appendix A.

4. Statistical press releases when data products are made available, produced, and issued by the statistical agency to provide a policy-neutral description of key findings and links to the data. Such releases must not include any policy commentary;[45]
5. Explanatory materials for all statistical product releases that assist users in understanding the product and convey the strengths and limitations of the data (see Practice 10); and
6. Archiving policies that guide decisions on which underlying data assets are to be retained, where they are to be archived (with the National Archives and Records Administration, or with an established archive maintained by an academic or other nonprofit institution or both), and how they are to be made accessible for future secondary analysis while protecting confidentiality.

Individual-level microdata files make it possible for users to conduct in-depth research and analyses that are not possible with aggregate data. Such files contain data for samples of individual respondents that have been processed to protect confidentiality by deleting, aggregating, or modifying any information that might permit individual identification (Federal Committee on Statistical Methodology, 2022). Statistical agencies should keep abreast of new developments in confidentiality protection so that they can continue to provide as much useful aggregate data and microdata as possible at a time of increasing threats to privacy and confidentiality. The increasing availability of the amount and type of data increases risk to confidentiality, even of previously released public-use data (NASEM, 2017a,b,c, 2024c). Statistical agencies must remain mindful that once released in public-use form, these microdata cannot be unreleased. Preparation of public-use data files must take these factors into account.

Restricted-Access Statistical Products

Some statistical agency data are deemed too difficult to protect in public releases and are made available to bona fide researchers only through some form of restricted access. To provide researchers with the ability to run their own analyses on restricted microdata, some statistical agencies provide access to an analysis engine on their websites that performs the selected statistical operations on the confidential data. Safeguards are built in so that the researcher cannot see the individual records and cannot obtain any output, such as too-detailed tabulations, that could identify individual respondents.[46]

---

[45]See *Statistical Policy Directive No. 4*, Section 6a, (OMB, 2008) in Appendix A.
[46]For example, https://nces.ed.gov/datalab/index.aspx

A second method, pioneered by NCES, is to grant licenses to individual researchers to analyze restricted microdata at their own sites for statistical purposes. Such licenses require that the researchers agree to follow strict procedures for protecting confidentiality and accept liability for penalties if confidentiality is breached.[47]

A third method is to allow approved researchers to analyze restricted microdata for statistical purposes at a secure physical site, such as one of the FSRDCs currently located at 33 universities and research organizations around the country.[48] The FSRDC network began as a Census Bureau initiative. It has now expanded as a shared service across the federal statistical system, directed through the ICSP through the FSRDC EXCOM subcommittee, with the Census Bureau serving as the program management office. The FSRDCs offer access to data from other participating agencies.[49]

A fourth method allows researchers to analyze restricted microdata for statistical purposes via a secure virtual site, such as virtual data enclaves and secure virtual desktop platforms. Some agencies, such as the Centers for Medicare & Medicaid Services, are moving away from data licenses toward secure virtual environments to support data access while protecting confidentiality.[50]

Recent Innovations in Facilitating Access to Confidential Data for Statistical Purposes

The Evidence Act is expected to facilitate greater use of data for statistical purposes. The Act requires OMB to issue guidance on tiers of access for data depending on their sensitivity and legal protections. The Act also required OMB to implement a single-application process for access to data from recognized statistical agencies and units.

The Standard Application Process

In December 2022, OMB issued M-23-04 Establishment of Standard Application Process Requirements on Recognized Statistical Agencies and Units. The goal of the SAP is to provide a standardized and transparent process for applying for access to confidential data.[51] The memorandum

---

[47] For example, https://nces.ed.gov/statprog/instruct.asp

[48] https://www.census.gov/about/adrm/fsrdc/locations.html#:~:text=There%20are%20currently%2033%20open,research%20institutions%2C%20and%20government%20agencies

[49] https://www.census.gov/fsrdc

[50] For example, https://coleridgeinitiative.org/adrf/ and https://www.norc.org/Research/Capabilities/Pages/data-enclave.aspX; https://www.icpsr.umich.edu/web/pages/appfed/index.html

[51] See also https://ncses.nsf.gov/initiatives/standard-application-process/annual-reports

describes the application process, and the roles assigned to responsible entities. (See also Appendix A.)

In brief, the SAP Governance Board, operating as a subcommittee of the ICSP, functions as an executive steering committee. The SAP Governance Board is responsible for setting criteria and approving requests from non-statistical agencies to access confidential statistical data (OMB, 2022a, p. 7). It is responsible for evaluating progress in meeting the objectives of the SAP and identifying opportunities for improvement (OMB, 2022a, p. 7).

Additionally, an SAP Program Management Office (PMO) "[...] is responsible for the development, operation, and maintenance of the SAP portal and any additional technical services required to facilitate the SAP" (OMB, 2022a, p. 7). The PMO is also responsible for managing the SAP data catalogue, documenting and managing implementation guidance, and supporting stakeholder engagement activities with the SAP Governance Board (OMB, 2022a, p. 7).

The access requirements described in M-23-04 describe tiered authorization requirements for access, although these are not tied to specific modes of access (OMB, 2022a, p. 19).

The National Secure Data Service Demonstration Project (NSDS-D)

There have been additional federal efforts to increase the means by which confidential data can be accessed for research purposes. The NSDS-D project was established by the 2022 CHIPS and Science Act.[52] The "goal of the NSDS-D project is to inform efforts for developing a shared services model that would streamline and innovate data sharing and linking to enable decision making at all levels of government and in all sectors."[53] The demonstration project will inform whether an NSDS will be established in the future. To that end, the NSDS-D will pilot potential services, technologies, techniques, and shared services that could be used within a potential NSDS. The focus is on novel research collaborations, data linkage methodologies, and privacy-preserving technologies and techniques. As of September 2024, 15 demonstration projects are active.

---

[52]A National Secure Data Service, or NSDS, was first introduced by the bipartisan Commission on Evidence-Based Policymaking in 2016, and later further envisioned through specific recommendations provided by the Advisory Committee on Data for Evidence Building in 2022.

[53]https://ncses.nsf.gov/about/national-secure-data-service-demo#card1893

America's DataHub Consortium

The National Science Foundation (NSF) and NCSES established America's DataHub Consortium (ADC)[54] in August 2021 through an NSF contract with Advanced Technology International. The ADC is a public-private partnership through which federal, state, and local government agencies can submit proposals and sponsor testing of data service structures and functions that, in turn, could inform the vision for the NSDS. Members of the ADC include small and large businesses, nonprofit organizations, and academic institutions.

The ADC and NSDS-D are distinct, but their efforts are complementary. The NSDS-D will create and test the infrastructure and resources that support this shared-service operational model. The ADC facilitates this evidence-building by streamlining collaboration between researchers, innovators, and subject matter experts.

PRACTICE 10: OPENNESS ABOUT SOURCES AND LIMITATIONS OF THE DATA PROVIDED

A statistical agency must be transparent about how it acquires data and produces statistics and must be open about the strengths and limitations of its data. No matter how high-quality the statistical data are, they contain some uncertainty and error. This does not mean the data are wrong but does mean that they need to be used with an understanding of their limitations. Statistical agencies need to communicate clearly to a wide range of potential users what the uncertainty in the data means for using the statistical information appropriately.

To be most effective, openness should be tailored to different user groups. For press releases disseminated to the public, the agency must make every effort to note both the meaning of the statistics and their limitations for various uses. For more technically trained users, detailed descriptions of methods and measures of quality should be made available (Federal Committee on Statistical Methodology, 2001, 2020; Mirel et al., 2023).

Openness requires that statistical releases from an agency include a full description of the purpose of the program; the methods and assumptions used for data collection, processing, and estimation; information about the quality and relevance of the data; analysis methods used; and the results of research on the methods and data (NASEM, 2019a, 2022d). Such transparency is essential for credibility with data users and stakeholders and for public trust. Thus, openness about statistical limitations requires much more than providing estimates of sampling error. In addition to a discussion

---

[54] https://www.americasdatahub.org/frequently-asked-questions/

of nonsampling errors (e.g., coverage errors, nonresponse errors, measurement errors, and processing errors), it is valuable to have a description of the concepts measured and how they relate to the major uses of the data (NASEM, 2022d). Descriptions of the shortcomings of the data should be provided in sufficient detail to permit a user to take them into account in analysis and interpretation. Descriptions of how the data relate to similar data collected by other agencies or generated in the private sector also should be provided, particularly when the estimates from two or more surveys or other data sources exhibit large differences that may have policy implications (NASEM, 2022d).

There is often a tension between timeliness and accuracy. When concerns for timeliness prompt the release of preliminary estimates (as is done for some economic indicators and has been done in response to COVID-19), consideration should be given to the frequency of revisions and the mode of presentation from the point of view of the users as well as the issuers of the data. Agencies that release preliminary estimates must educate the public about differences among preliminary, revised, and final estimates.

An important aspect of openness concerns the treatment of errors that are discovered subsequent to data release. Openness means that an agency has an obligation to issue corrections publicly and in a timely manner. The agency should use not only the same dissemination avenues to announce corrections that it used to release the original statistics, but also additional vehicles, as appropriate, to alert the widest possible audience of current and future users of the corrections in the information. Agencies should be proactive in seeking ways to alert potential users of the data about the nature of a problem and the corrective actions that it is taking or that users should take.

Federal statistical agencies should implement quality frameworks for their programs and use these to describe the strengths and limitations of the statistical information produced by the data (see Practices 3 and 6). Some statistical agencies have developed "quality profiles" for major surveys, which document what is and is not known about errors in estimates to help users (NASEM, 2022d).

In cases where agencies are using data from administrative and other alternative sources, they need to provide information not only on what is known about those sources but also on how the data were linked or blended with other data sources and the potential errors introduced through linkage. This is a challenging and evolving area (NASEM, 2017a), and efforts are ongoing[55] to develop best practices and quality frameworks for these topics (Czajka & Strange, 2018; Federal Committee on Statistical Methodology, 2020; NASEM, 2022d, 2023b,c, 2024c; Prell et al., 2019).

---

[55] https://www.fcsm.gov/groups/data-quality/

Statistical agencies should treat the effort to provide information on the quality, limitations, and appropriate use of their data as an essential part of their mission. Such information and metadata should be readily accessible to all known and potential users (NASEM, 2022d; NRC, 1993b, 1997a, 2007a). By being open about the sources and limitations of their data and by providing as much data as possible in ways that are as easy as possible for users to access and apply, federal statistical agencies fulfill their vital mission to inform the public, contribute to evidence-based policymaking, and support the development of societal knowledge for the public good.

Acronyms and Abbreviations

ACS   American Community Survey
ADC   America's DataHub Consortium
APHIS   Animal and Plant Health Inspection Service (of Department of Agriculture)

BEA   Bureau of Economic Analysis (of Department of Commerce)
BJS   Bureau of Justice Statistics (of Department of Justice)
BLS   Bureau of Labor Statistics (of Department of Labor)
BTS   Bureau of Transportation Statistics (of Department of Transportation)

CAIO   Chief Artificial Intelligence Officer
CBHSQ   Center for Behavioral Health Statistics and Quality (of Department of Health and Human Services)
Census Bureau of the Census (of Department of Commerce)
CEP   U.S. Commission on Evidence-Based Policymaking
CFO   Chief Financial Officer
CIPSEA Confidential Information Protection and Statistical Efficiency Act
CNSTAT Committee on National Statistics
CPI   Consumer Price Index
CSOTUS Chief Statistician of the U.S.

DHS   Department of Homeland Security
DOD   Department of Defense

DOE	Department of Energy
DOI	Department of Interior
EIA	Energy Information Administration (of Department of Energy)
EOP	Executive Office of the President
EPA	Environmental Protection Agency
ERS	Economic Research Service (of Department of Agriculture)
FCSM	Federal Committee on Statistical Methodology
FRB	Federal Reserve Board
FSRDC	Federal Statistical Research Data Center
GSA	General Services Administration
HHS	Department of Health and Human Services
HUD	Department of Housing and Urban Development
ICSP	Interagency Council on Statistical Policy
IRB	institutional review board
IRS	Internal Revenue Service
MSS	Microeconomic Surveys Section (of the Federal Reserve Board)
NAICS	North American Industry Classification System
NASA	National Aeronautics and Space Administration
NASS	National Agricultural Statistical Service (of Department of Agriculture)
NCES	National Center for Education Statistics (of Department of Education)
NCHS	National Center for Health Statistics (of Department of Health and Human Services)
NCSES	National Center for Science and Engineering Statistics (of National Science Foundation)
NRC	National Research Council
NRC	Nuclear Regulatory Commission
NSDS	National Secure Data Service
NSF	National Science Foundation
OHSS	Office of Homeland Security Statistics (of Department of Homeland Security)
OMB	Office of Management and Budget (of Executive Office of the President)
OPM	Office of Personal Management

ORES	Office of Research, Evaluation, and Statistics (of Social Security Administration)
OSTP	Office of Science and Technology Policy (of Executive Office of the President)
PD&R	Office of Policy Development and Research (of Department of Housing and Urban Development)
SAMHSA	Substance Abuse and Mental Health Services Administration (of Department of Health and Human Resources)
SAP	Standard Application Process
SBA	Small Business Administration
SOI	Statistics of Income (of Department of the Treasury)
SSP	Statistical and Science Policy Office (of Office of Management and Budget); also known as the Office of the Chief Statistician of the United States
State	Department of State
Treasury	Department of the Treasury
USAID	U.S. Agency for International Development
USDA	Department of Agriculture
VA	Department of Veterans Affairs

Glossary of Selected Terms

Advisory Committee on Data for Evidence Building
   Established as part of the Foundations for Evidence-Based Policymaking Act of 2018 (Foundations for Evidence-Based Policymaking Act of 2018, 2019) to review, analyze, and make recommendations to the director of the White House Office of Management and Budget (OMB) on how to promote the use of federal data for evidence building (Advisory Committee on Data for Evidence Building, 2021).

Chief Artificial Intelligence Officers
   As required by Executive Order 14110, each agency must designate a Chief Artificial Intelligence Officer (CAIO). See M-24-10 for a detailed description of roles, responsibilities, seniority, position, and reporting structure for agency CAIOs. "At CFO Act agencies, a primary role of the CAIO must be coordination, innovation, and risk management for their agency's use of AI specifically, as opposed to data or IT issues in general" (OMB, 2024f, p. 5).

Chief Data Officers
   Chief Data Officers are "designated on the basis of demonstrated training and experience in data management, governance (including creation, application, and maintenance of data standards), collection, analysis, protection, use, and dissemination, including with respect to any statistical and related techniques to protect and de-identify confidential data"; their responsibilities include "be[ing] responsible for lifecycle data management; coordinat[ing] with any official in the agency responsible for

using, protecting, disseminating, and generating data to ensure that the data needs of the agency are met; manag[ing] data assets of the agency, including the standardization of data format, sharing of data assets, and publication of data assets in accordance with applicable law; ensur[ing] that, to the extent practicable, agency data conforms with data management best practices; and engag[ing] agency employees, the public, and contractors in using public data assets and encourage[ing] collaborative approaches on improving data use" (Foundations for Evidence-Based Policymaking Act of 2018, 2019 § 3520).

CFO Act agencies
The Chief Financial Officer (CFO) Act of 1990 gave OMB new authority and responsibility for directing federal financial management, modernizing the government's financial management systems, and strengthening financial reporting. The 24 CFO Act agencies include: Agency for International Development*; Department of Agriculture; Department of Commerce; Department of Defense*; Department of Education; Department of Energy; Department of Health and Human Services; Department of Homeland Security*; Department of Housing and Urban Development*; Department of the Interior*; Department of Justice; Department of Labor; Department of State*; Department of Transportation; Department of the Treasury; Department of Veterans Affairs*; Environmental Protection Agency*; General Services Administration*; National Aeronautics and Space Administration*; National Science Foundation; Nuclear Regulatory Commission*; Office of Personnel Management*; Small Business Administration*; and Social Security Administration. The Evidence Act added representation of all CFO Act agencies to the Interagency Council on Statistical Policy (ICSP) and created the role of Statistical Official in each of these agencies where a statistical agency was not already present (denoted here by *).[1]

CIPSEA
The Confidential Information Protection and Statistical Efficiency Act (CIPSEA), initially adopted in 2002, was reauthorized and expanded as Title III of the Evidence Act (see below; Foundations for Evidence-Based Policymaking Act of 2018, 2019).

Confidentiality
"Data or information acquired by an agency under a pledge of confidentiality and for exclusively statistical purposes shall be used by officers, employees, or agents of the agency exclusively for statistical purposes"

---

[1] https://www.cio.gov/handbook/it-laws/cfo-act/

(Confidential Information Protection and Statistical Efficiency Act, 2002 § 102(a)).

Data holders
Organizations that hold information of possible use in a national data infrastructure. These include federal, state, local, and tribal agencies, as well as other public and private-sector organizations (National Academies of Sciences, Engineering, and Medicine [NASEM], 2023b).

Data infrastructure
This term includes data assets; the technologies used to discover, access, share, process, use, analyze, manage, store, preserve, protect, and secure those assets; the people, capacity, and expertise needed to manage, use, interpret, and understand data; the guidance, standards, policies, and rules that govern data access, use, and protection; the organizations and entities that manage, oversee, and govern the data infrastructure; and the communities and data subjects whose data is shared and used for statistical purposes and may be impacted by decisions that are made using those data assets (NASEM, 2023d).

Data subjects
The people, entities, or organizations described by data files (NASEM, 2023b). In the prior edition of this publication, these were referred to as "data providers." Increasingly in new sources of data, the people, entities, and organizations described by the data do not directly provide the data themselves. Instead, these data may come from data holders (see *data holders*).

Evaluation Officers
Evaluation Officers are "designated without regard to political affiliation and based on demonstrated expertise in evaluation methodology and practices and appropriate expertise to the disciplines of the agency;" their responsibilities include "continually assess[ing] the coverage, quality, methods, consistency, effectiveness, independence, and balance of the portfolio of evaluations, policy research, and ongoing evaluation activities of the agency; assess[ing] agency capacity to support the development and use of evaluation; and establish[ing] and implement[ing] an agency evaluation policy" (Foundations for Evidence-Based Policymaking Act of 2018, 2019 § 313).

Evidence Act
Also referred to as the Foundations for Evidence-Based Policymaking Act of 2018. This law requires agency data to be accessible and requires

agencies to plan to develop statistical evidence to support policymaking (Foundations for Evidence-Based Policymaking Act of 2018, 2019). Title II of the Evidence Act is referred to as the OPEN Government Data Act (see below). Title III of the Evidence Act is referred to as CIPSEA 2018 (see above).

Federal Committee on Statistical Methodology (FCSM)
"Founded in 1975 by the Statistical and Science Policy Branch (formerly, the Office of Statistical Policy) in the Office of Information and Regulatory Affairs in the Office of Management and Budget (OMB), the FCSM assists in carrying out SSP/OMB's role in setting and coordinating statistical policy. The FCSM serves as a resource for OMB and the federal statistical system to inform decision making on matters of statistical policy and to provide technical assistance and guidance on statistical and methodological issues."[2] Members of FCSM are federal staff who contribute their professional expertise to advancing the coordination of federal statistical policy in addition to meeting their leadership responsibilities at their respective ICSP agencies.

Federal statistical system
The federal statistical system comprises the Office of the U.S. Chief Statistician, recognized federal statistical agencies, recognized federal statistical units, federal Statistical Officials in other CFO agencies, and over 100 other federal statistical programs. It is supported by state, local, and other federal agencies that provide data to produce federal statistics and the groups that provide guidance to connect the system, including the FCSM.

Interagency Council on Statistical Policy (ICSP)
"Formed in 1989, the ICSP was created to improve communication among the heads of the principal statistical agencies, and later was charged with advising and assisting the CSOTUS [Chief Statistician of the United States]. The Evidence Act expanded membership to include the newly established Statistical Officials across major cabinet agencies, 11 of which are also heads of recognized statistical agencies. Led by the CSOTUS, the ICSP supports implementation of the statistical system's vision to operate as a seamless system. The 30 members of ICSP include 13 recognized statistical agencies, 3 recognized statistical units, 13 Statistical Officials of other CFO agencies (those not with a statistical agency), and the CSCOTUS."[3]

---

[2]https://www.fcsm.gov/about/
[3]https://www.statspolicy.gov/about/#icsp

OPEN Government Data Act
Title II of the Evidence Act (see above; Foundations for Evidence-Based Policymaking Act of 2018, 2019).

Privacy
"Privacy refers to freedom from intrusion into one's personal matters and personal information" (NASEM, 2023d, p. 75).

Recognized statistical agencies
In this report, the descriptors "principal statistical agencies" and "recognized statistical agencies" are used interchangeably. The U.S. recognized federal statistical agencies are (13): Bureau of Economic Analysis (Department of Commerce); Bureau of Justice Statistics (Department of Justice); Bureau of Labor Statistics (Department of Labor); Bureau of Transportation Statistics (Department of Transportation); Census Bureau (Department of Commerce); Economic Research Service (Department of Agriculture); Energy Information Agency (Department of Energy); National Agricultural Statistics Service (Department of Agriculture); National Center for Education Statistics (Department of Education); National Center for Health Statistics (Department of Health and Human Services); National Center for Science and Engineering Statistics (National Science Foundation); Office of Research, Evaluation, and Statistics (Social Security Administration); and Statistics of Income (Department of the Treasury; Foundations for Evidence-Based Policymaking Act of 2018, 2019).

Recognized statistical units
In this report, the descriptors "designated statistical units" and "recognized statistical units" are used interchangeably, reflecting differences in language over time. Recognized statistical units are organizational units embedded in a nonstatistical agency that produce and disseminate statistics under a pledge of confidentiality under the requirements of CIPSEA. There are three recognized federal statistical units: Microeconomic Surveys Unit (Federal Reserve Board); Center for Behavioral Health Statistics and Quality (Substance Abuse and Mental Health Services Administration; Department of Health and Human Services); and National Animal Health Monitoring System (Animal and Plant Health Inspection Service, Department of Agriculture; Foundations for Evidence-Based Policymaking Act of 2018, 2019).

Standard Application Process (SAP)
A uniform method for accessing federal confidential data assets. When fully built, the SAP will serve as a "front door" through which to

apply for permission to use protected data from any of the 16 federal statistical agencies and designated units for evidence building[4] (Office of Management and Budget, 2022a). The current portal is at: https://www.researchdatagov.org

Statistical activities
"[... T]he collection, compilation, processing, or analysis of data for the purpose of describing or making estimates concerning the whole, or relevant groups or components within, the economy, society, or the natural environment; and [...] includes the development of methods or resources that support those activities, such as measurement methods, models, statistical classifications, or sampling frames" (Confidential Information Protection and Statistical Efficiency Act, 2002 § 2 (7)).

Statistical agency or unit
"[... A]n agency or organizational unit of the executive branch whose activities are predominantly the collection, compilation, processing, or analysis of information for statistical purposes" (Confidential Information Protection and Statistical Efficiency Act, 2002 § 2 (8)).

Statistical Official
"The head of each [Chief Financial Officer Act] agency shall designate the head of any statistical agency or unit within the agency, or in the case of an agency that does not have a statistical agency or unit, any senior agency official with appropriate expertise, as a statistical official to advise on statistical policy, techniques, and procedures. Agency officials engaged in statistical activities may consult with any such statistical official as necessary" (Foundations for Evidence-Based Policymaking Act of 2018, 2019 § 314). At the 11 CFO Act agencies that contain a recognized statistical agency or unit (RSAU), the head of that RSAU has been designated the Statistical Official, as required by M-19-23.

Statistical purpose
"[... T]he description, estimation, or analysis of the characteristics of groups, without identifying the individuals or organizations that comprise such groups; and...includes the development, implementation, or maintenance of methods, technical or administrative procedures, or information resources that support the purposes" (Confidential Information Protection and Statistical Efficiency Act, 2002 § 2 (9)).

---

[4] https://ncses.nsf.gov/about/standard-application-process

Tiered access
"An application of data minimization, a key privacy safeguard for evidence building [...] Data minimization means giving access to the least amount of data needed to complete an approved project. Tiered access uses a variety of controls to minimize data needed for a given project and thereby reduce risk of disclosure of confidential data" (Commission on Evidence-Based Policymaking, 2017, p. 38).

U.S. Commission on Evidence-Based Policymaking
The Evidence-Based Policymaking Commission Act of 2016 created the bipartisan Commission on Evidence-Based Policymaking. The 15-member Commission was charged with examining all aspects of how to increase the availability and use of government data to build evidence and inform program design, while protecting the privacy and confidentiality of those data (Evidence-Based Policymaking Commission Act of 2016, 2016). The Commission's report (Commission on Evidence-Based Policymaking, 2017) informed the Evidence Act (Foundations for Evidence-Based Policymaking Act of 2018, 2019).

References

Abascal, M. A., Archie, J. C., Crawford, G. E., Naftalis, B. A., Schindler, D. J., Jones, S. C., & Stout, A. L. (2016). *What you need to know about the Cybersecurity Act of 2015*. Latham & Watkins. https://www.lw.com/thoughtLeadership/lw-Cybersecurity-Act-of-2015

Advisory Committee on Data for Evidence Building. (2021). *Advisory Committee on Data for Evidence Building: Year 1 report*. https://www.bea.gov/system/files/2021-10/acdeb-year-1-report.pdf

Advisory Committee on Data for Evidence Building. (2022). *Advisory Committee on Data for Evidence Building: Year 2 report*. https://www.bea.gov/system/files/2022-10/acdeb-year-2-report.pdf

American Economic Association. (2018). *Principles of economic measurement*. https://www.aeaweb.org/content/file?id=6847

American Statistical Association. (2022). *Ethical guidelines for statistical practice*. https://www.amstat.org/your-career/ethical-guidelines-for-statistical-practice

American Statistical Association. (2024). *The nation's data at risk: Meeting America's information needs for the 21st century*. https://www.amstat.org/policy-and-advocacy/the-nation's-data-at-risk-meeting-american's-information-needs-for-the-21st-century

An Act to Establish a Department of Education, Pub. Law 39-73, 14 Stat. 434 (1867). https://www.docsteach.org/documents/document/act-of-march-2-1867-public-law-3973-14-stat-434-which-established-the-department-of-education

Anderson, M. J. (2015). *The American census: A social history* (2nd ed.). Yale University Press. https://doi.org/10.12987/9780300216967

Behzad, B., Bheem, B., Elizondo, D., & Martonosi, S. (2023). Prevalence and propagation of fake news. *Statistics and Public Policy*, *10*(1). https://doi.org/10.1080/2330443X.2023.2190368

Bohman, M. (2024). Making a difference through trusted, high-quality research and statistics. *American Journal of Agricultural Economics*, *106*(2), 485–495. https://doi.org/https://doi.org/10.1111/ajae.12459

Bowen, C., & Snoke, J. (2023). *Do no harm guide: Applying equity awareness in data privacy methods*. Urban Institute. https://policycommons.net/artifacts/3525920/do-no-harm-guide/4326655/

Bowen, C. M. (2023). The autonomy gap: Response to Citro et al. and the statistical community. *Statistics and Public Policy, 10*(1). https://doi.org/10.1080/2330443X.2023.2221324

Budget and Accounting Procedures Act, 31 U.S.C. § 1104(d) (1950). https://www.govregs.com/uscode/expand/title31_subtitleII_chapter11_section1104#uscode_3

Bureau of Labor Statistics (BLS). (2021a). *Final report of the Interagency Technical Working Group on evaluating alternative measures of poverty*. U.S. Department of Labor. https://www.bls.gov/cex/itwg-report.pdf

BLS. (2021b). *Report to the Office of Management and Budget: Consumer inflation measures*. U.S. Department of Labor. https://www.bls.gov/evaluation/technical-recommendations-for-the-consumer-inflation-measure-best-suited-for-conducting-annual-adjustments-to-the-official-poverty-measure.pdf#:~:text=The%20ITWG%20was%20chartered%20to%3A%20%281%29%20develop%20a,the%20methodologies%20used%20in%20the%20consumer%20inflation%20measures

Cabrera, N. B., Bongartz, E. C., Dörrenbächer, N., Goebel, J., Kaluza, H., & Siegers, P. (2020). White paper on implementing the FAIR principles for data in the social, behavioural, and economic sciences. *Consortium for the Social, Behavioral, Educational, and Economic Sciences, 274*. https://doi.org/10.17620/02671.60

Census Bureau. (2010). *Observations from the Interagency Technical Working Group on Developing a Supplemental Poverty Measure*. U.S. Department of Commerce. https://www.census.gov/content/dam/Census/library/working-papers/2010/demo/SPM_Wkg-Grp.pdf

Census Bureau. (2021). *Improvements to the Census Bureau's supplemental poverty measure for 2021*. U.S. Department of Commerce. https://www.census.gov/topics/income-poverty/supplemental-poverty-measure/library/working-papers/topics/potential-changes.html

Chief Financial Officers Act of 1990, Pub. L. No. 101-576, 105 Stat. 2838 (1990). https://www.congress.gov/101/statute/STATUTE-104/STATUTE-104-Pg2838.pdf

CHIPS and Science Act of 2022, Pub. L. No. 117-167, 136 Stat. 1366 (2022). https://www.congress.gov/bill/117th-congress/house-bill/4346/text

Citro, C. F. (2014). Principles and practices for a federal statistical agency: Why, what, and to what effect. *Statistics and Public Policy, 1*(1), 51–59. http://dx.doi.org/10.1080/2330443X.2014.912953

Citro, C. F. (2016). The U.S. federal statistical system's past, present, and future. *Annual Review of Statistics and Its Application, 3*(1), 347–373. https://www.annualreviews.org/content/journals/10.1146/annurev-statistics-041715-033405

Citro, C. F., Auerbach, J., Evans, K. S., Groshen, E. L., Landefeld, J. S., Mulrow, J., Petska, T., Pierson, S., Potok, N., Rothwell, C. J., Thompson, J., Woodworth, J. L., & Wu, E. (2023). What protects the autonomy of the federal statistical agencies? An assessment of the procedures in place to protect the independence and objectivity of official U.S. statistics. *Statistics and Public Policy, 10*(1). https://doi.org/10.1080/2330443X.2023.2188062

Cohen, M. (2023). Discussion of "What protects the autonomy of the federal statistical agencies? An assessment of the procedures in place to protect the independence and objectivity of official U.S. statistics" by Citro et al. (2023). *Statistics and Public Policy, 10*(1). https://doi.org/10.1080/2330443X.2023.2244026

Commission on Evidence-Based Policymaking. (2017). *The promise of evidence-based policymaking*. https://bipartisanpolicy.org/wp-content/uploads/2019/03/Full-Report-The-Promise-of-Evidence-Based-Policymaking-Report-of-the-Comission-on-Evidence-based-Policymaking.pdf

REFERENCES

Committee on National Statistics. (2024). *People and publications: 1972–2023.* Division of Behavioral and Social Sciences and Education. https://www.nationalacademies.org/documents/embed/link/LF2255DA3DD1C41C0A42D3BEF0989ACAECE3053A6A9B/file/D48F08611AA79D848FF6C76DDB21DF2737D47EE3883B?noSaveAs=1

Confidential Information and Statistical Efficiency Act of 2018, Pub. L. No. 115-435, 132 Stat. 5529. (2019). https://www.congress.gov/115/plaws/publ435/PLAW-115publ435.pdf

Confidential Information Protection and Statistical Efficiency Act, Pub. L. No. 107-347, H.R. 5215. (2002). https://www.congress.gov/107/bills/hr5215/BILLS-107hr5215rh.pdf

Cui, I., Ho, D. E., Martin, O., & O'Connell, A. J. (2024, forthcoming). *Governing by assignment.* Stanford Law School. https://dho.stanford.edu/wp-content/uploads/IPA.pdf

Cybersecurity & Infrastructure Security Agency. (2023). *Einstein* [archived page]. U.S. Department of Homeland Security. https://web.archive.org/web/20230716151116/https:/www.cisa.gov/einstein

Cybersecurity Information Sharing Act, Pub. Law 114-185, 130 Stat. 538 (2016). https://www.congress.gov/114/plaws/publ185/PLAW-114publ185.pdf

Czajka, J., & Strange, M. (2018). *Transparency in the reporting of quality for integrated data: A review of international standards and guidelines.* U.S Department of Commerce. https://www.washstat.org/presentations/20180226/20180226_Czajka.pdf#:~:text=%E2%80%A2%20Quality%20reporting%20on%20accuracy%20is

Department of Commerce. (1978a). *43 F.R. 19308.* https://www.govinfo.gov/content/pkg/FR-1978-05-04/pdf/FR-1978-05-04.pdf

Department of Commerce. (1978b). *43 F.R. 19260.* https://www.govinfo.gov/content/pkg/FR-1978-05-04/pdf/FR-1978-05-04.pdf

Department of Commerce. (2014). *Fostering innovation, creating jobs, driving better decisions: The value of government data.* https://www.commerce.gov/data-and-reports/reports/2014/07/fostering-innovation-creating-jobs-driving-better-decisions-value#:~:text=This%20report%20finds:%20Government%20data%20potentially

Duncan, J. W., & Shelton, W. C. (1978). *Revolution in United States statistics, 1926–1976.* U.S. Government Printing Office. https://babel.hathitrust.org/cgi/pt?id=uiug.30112052125652&seq=185

E-Government Act, Pub. Law 107-347, 116 Stat. 2899 (2002). https://www.congress.gov/107/plaws/publ347/PLAW-107publ347.pdf

European Statistical System Committee. (2017). *European statistics code of practice.* Publications Office of the European Union. https://ec.europa.eu/eurostat/documents/4031688/8971242/KS-02-18-142-EN-N.pdf/e7f85f07-91db-4312-8118-f729c75878c7

Evidence-Based Policymaking Commission Act of 2016, Pub. L. No. 114-140, 130 Stat. 317 (2016). https://www.govinfo.gov/content/pkg/STATUTE-130/pdf/STATUTE-130-Pg317.pdf

Excepted Service, 5 U.S.C. 2103 (1978). https://www.govinfo.gov/content/pkg/USCODE-2023-title5/pdf/USCODE-2023-title5-partIII-subpartA-chap21-sec2103.pdf

Executive Office of the President. (1933). *Executive Order 6226: Providing for current encumbrance reports.* https://www.presidency.ucsb.edu/documents/executive-order-6226-providing-for-current-encumbrance-reports

Executive Office of the President. (1977). *Executive Order 12013: Statistical policy functions.* https://www.presidency.ucsb.edu/documents/executive-order-12013-statistical-policy-functions

Executive Office of the President. (1981). *Executive Order 12318: Statistical policy functions.* https://www.presidency.ucsb.edu/documents/executive-order-12318-statistical-policy-functions

Executive Office of the President. (2009). *M 3-9-09: Scientific integrity.* https://obamawhitehouse.archives.gov/the-press-office/memorandum-heads-executive-departments-and-agencies-3-9-09

Executive Office of the President. (2023). *Executive Order 14110: Safe, secure, and trustworthy development and use of artificial intelligence.* https://www.govinfo.gov/content/pkg/FR-2023-11-01/pdf/2023-24283.pdf

Federal Committee on Statistical Methodology. (2001). *Statistical policy working paper No. 31: Measuring and reporting sources of error in surveys.* https://nces.ed.gov/FCSM/pdf/spwp31.pdf

Federal Committee on Statistical Methodology. (2020). *A framework for data quality.* https://nces.ed.gov/FCSM/pdf/FCSM.20.04_A_Framework_for_Data_Quality.pdf

Federal Committee on Statistical Methodology. (2022). *Data protection toolkit: Report and resources on statistical disclosure limitation methodology and tiered data access (formerly 'Statistical Policy Working Paper No. 22').* https://nces.ed.gov/fcsm/dpt

Federal Committee on Statistical Methodology. (2024). U.S. Office of Management and Budget. https://www.fcsm.gov/

Federal Information Technology Acquitision Reform Act, Pub. Law 113-291, 128 Stat. 3438 (2014). https://www.cio.gov/assets/files/FITARA%20Pub%20L%20113-291.pdf

Federal Interagency Forum on Aging-Related Statistics. (2024). https://agingstats.gov/

Federal Interagency Forum on Child and Family Statistics. (2024). https://www.childstats.gov/

Federal Policy for the Protection of Human Subjects, 45 F.R. Part 46(a). (2018). https://www.ecfr.gov/on/2018-07-19/title-45/subtitle-A/subchapter-A/part-46

Federal Reports Act, Pub. L. No. 831, 31–33 (1942). https://www.census.gov/history/pdf/Federal_Reports_Act_1942.pdf

Foundations for Evidence-Based Policymaking Act of 2018, Pub. L. No. 115-435, 132 Stat. 5529. (2019). https://www.govinfo.gov/content/pkg/PLAW-115publ435/html/PLAW-115publ435.htm

General Services Administration. (2020). *Federal data strategy: Data ethics framework.* https://resources.data.gov/assets/documents/fds-data-ethics-framework.pdf

Goerge, R. (2018). Barriers to accessing state data and approaches to addressing them. *Annals of the American Academy of Political and Social Science, 675,* 122–137. https://journals.sagepub.com/doi/pdf/10.1177/0002716217741257

Gotterbarn, D., Bruckman, A., Flick, C., Miller, K., & Wolf, M. J. (2018). ACM code of ethics: A guide for positive action. *Communications of the Association for Computing Machinery, 61*(1), 121–128. https://doi.org/10.1145/3173016

Government Accountability Office (GAO). (1995). *Report No. GGD-95-65: Statistical Agencies: Adherence to guidelines and coordination of budgets.* http://www.gao.gov/products/GGD-95-65

GAO. (2007). *Bureau of Justice Statistics Report No. GAO-07-340: Quality guidelines generally followed for police-public contact surveys, but opportunities exist to help assure agency independence.* http://www.gao.gov/products/GAO-07-340

GAO. (2012). *Report No. GAO-12-54: Federal statistical system: Agencies can make greater use of existing data, but continued progress is needed on access and quality issues.* http://www.gao.gov/products/GAO-12-54

Gravelle, H., & Rees, R. (2004). *Microeconomics* (3rd ed.). Pearson Education Limited. https://pure.york.ac.uk/portal/en/publications/microeconomics

Group of Eight. (2013). *G8 Open Data Charter.* https://www.gov.uk/government/publications/open-data-charter/g8-open-data-charter-and-technical-annex

Habermann, H., & Louis, T. A. (2020). Can the fundamental principles of official statistics and the political process co-exist? *Statistical Journal of the IAOS*, 36(2), 347–353. https://content.iospress.com/download/statistical-journal-of-the-iaos/sji200624?id=statistical-journal-of-the-iaos%2Fsji200624

Habermann, H., Louis, T. A., & Reeder, F. (2023). Is autonomy possible and is it a good thing? *Statistics and Public Policy*, 10(1). https://doi.org/10.1080/2330443X.2023.2221314

Hartman, K., Habermann, H., Harris-Kojetin, B., Jones, C., & Louis, T. (2014). Strength under pressure. *Significance*, 11(4), 44–47. https://click.endnote.com/viewer?doi=10.1111/J.1740-9713.2014.00769.X&route=2

Health Insurance Portability and Accountability Act, Pub. Law 104-191, 110 Stat. 1936 (1996). https://www.congress.gov/104/plaws/publ191/PLAW-104publ191.pdf

Hendriks, C. (2012). *Input data quality in register based statistics: The Norwegian experience.* American Statistical Association. https://ww2.amstat.org/meetings/proceedings/2012/data/assets/pdf/303710_71783.pdf

Hillygus, D. S., Nie, N. H., Prewitt, K., & Pals, H. (2006). *The hard count: The political and social challenges of Census mobilization.* Russell Sage Foundation. http://www.jstor.org/stable/10.7758/9781610442886

Ho, D. E., & O'Connell, A. J. (2024). Opinion: The government has a workforce crisis. One of its best-kept secrets can fix it. *Washington Post.* https://www.washingtonpost.com/opinions/2024/03/12/federal-government-workforce-crisis/

Hogan, H., & Steffey, D. (2014). *Professional ethics for statisticians: An organizational history.* https://ww2.amstat.org/meetings/proceedings/2014/data/assets/pdf/311653_87910.pdf

Hughes-Cromwick, E., & Coronado, J. (2019). The value of US government data to US business decisions. *Journal of Economic Perspectives*, 33(1), 131–146. https://www.aeaweb.org/articles?id=10.1257/jep.33.1.131

Information Quality Act, Pub. L. No. 106-554, 114 Stat. 2763A § 515. (2000). https://www.govinfo.gov/content/pkg/PLAW-106publ554/pdf/PLAW-106publ554.pdf

Intergovernmental Personnel Act of 1970, Pub. L. No. 91-648. (1970). https://uscode.house.gov/statutes/pl/91/648.pdf

Internal Revenue Code of 1986, Pub. L. No. 99-514, 100 Stat. 2095 § 2 § 6103. (1986). https://www.govinfo.gov/content/pkg/USCODE-2011-title26/pdf/USCODE-2011-title26-subtitleF-chap61-subchapB-sec6103.pdf

Martin, M. E. (1981). Statistical practice in bureaucracies. *Journal of the American Statistical Association*, 76(373), 1–8. http://www.tandfonline.com/doi/abs/10.1080/01621459.1981.10477593

Metropolitan Areas Protection and Standardization Act of 2021, Pub. L. No. 117-219, 136 Stat. 2271. (2021). https://www.congress.gov/117/plaws/publ219/PLAW-117publ219.pdf

Mirel, L. B., Singpurwalla, D., Hoppe, T., Liliedahl, E., Schmitt, R., & Weber, J. (2023). *A framework for data quality: Case studies.* Federal Committee on Statistical Methodology. https://www.fcsm.gov/assets/files/docs/FCSM.23.02_DQ_case_studies_FINAL.pdf

National Academies of Sciences, Engineering, and Medicine (NASEM). (2016a). *Reducing respondent burden in the American Community Survey: Proceedings of a workshop.* The National Academies Press. https://doi.org/10.17226/23639

NASEM. (2016b). *Making eye health a population health imperative: Vision for tomorrow.* The National Academies Press. https://doi.org/10.17226/23471

NASEM. (2017a). *Innovations in federal statistics: Combining data sources while protecting privacy.* The National Academies Press. https://doi.org/10.17226/24652

NASEM. (2017b). *Improving crop estimates by integrating multiple data sources.* The National Academies Press. https://doi.org/10.17226/24892

NASEM. (2017c). *Federal statistics, multiple data sources, and privacy protection: Next steps.* The National Academies Press. https://doi.org/10.17226/24893

NASEM. (2017d). *Principles and practices for federal program evaluation: Proceedings of a workshop.* The National Academies Press. https://doi.org/10.17226/24831

NASEM. (2017e). *Advancing concepts and models for measuring innovation: Proceedings of a workshop.* The National Academies Press. https://doi.org/10.17226/23640

NASEM. (2018a). *The 2014 redesign of the survey of income and program participation: An assessment.* The National Academies Press. https://doi.org/10.17226/24864

NASEM. (2018b). *Measuring the 21st century science and engineering workforce population: Evolving needs.* The National Academies Press. https://doi.org/10.17226/24968

NASEM. (2018c). *Reengineering the Census Bureau's annual economic surveys.* The National Academies Press. https://doi.org/10.17226/25098

NASEM. (2019a). *Methods to foster transparency and reproducibility of federal statistics: Proceedings of a workshop.* The National Academies Press. https://doi.org/10.17226/25305

NASEM. (2019b). *Improving the American Community Survey: Proceedings of a workshop.* The National Academies Press. https://doi.org/10.17226/25387

NASEM. (2019c). *Improving data collection and measurement of complex farms.* The National Academies Press. https://doi.org/10.17226/25260

NASEM. (2020a). *Measuring alternative work arrangements for research and policy.* The National Academies Press. https://doi.org/10.17226/25822

NASEM. (2020b). *A consumer food data system for 2030 and beyond (prepublication).* The National Academies Press. https://doi.org/10.17226/25657

NASEM. (2021a). *Principles and practices for a federal statistical agency: Seventh edition.* The National Academies Press. https://doi.org/10.17226/25885

NASEM. (2021b). *A satellite account to measure the retail transformation: Organizational, conceptual, and data foundations.* The National Academies Press. https://doi.org/10.17226/26101

NASEM. (2022a). *A vision and roadmap for education statistics.* The National Academies Press. https://doi.org/10.17226/26392

NASEM. (2022b). *Modernizing the consumer price index for the 21st century.* The National Academies Press. https://doi.org/10.17226/26485

NASEM. (2022c). *Measuring sex, gender identity, and sexual orientation.* The National Academies Press. https://doi.org/10.17226/26424

NASEM. (2022d). *Transparency in statistical information for the National Center for Science and Engineering Statistics and all federal statistical agencies.* The National Academies Press. https://doi.org/10.17226/26360

NASEM. (2023a). *A roadmap for disclosure avoidance in the Survey of Income and Program Participation.* The National Academies Press. https://doi.org/10.17226/27169

NASEM. (2023b). *Toward a 21st century national data infrastructure: Mobilizing information for the common good.* The National Academies Press. https://nap.nationalacademies.org/read/26688/chapter/1

NASEM. (2023c). *Toward a 21st century national data infrastructure: Enhancing survey programs by using multiple data sources.* The National Academies Press. https://doi.org/10.17226/26804

NASEM. (2023d). *2020 Census data products: Demographic and housing characteristics file: Proceedings of a workshop.* The National Academies Press. https://doi.org/10.17226/26727

NASEM. (2023e). *An updated measure of poverty: (Re)drawing the line.* The National Academies Press. https://doi.org/10.17226/26825

NASEM. (2024a). *Reducing intergenerational poverty.* The National Academies Press. https://doi.org/10.17226/27058

NASEM. (2024b). *An integrated system of U.S. household income, wealth, and consumption data and statistics to inform policy and research.* The National Academies Press. https://doi.org/10.17226/27333

NASEM. (2024c). *Toward a 21st century national data infrastructure: Managing privacy and confidentiality risks with blended data*. The National Academies Press. https://doi.org/10.17226/27335

National Artificial Intelligence Initiative Act, H.R. 6216. (2020). https://www.congress.gov/bill/116th-congress/house-bill/6216

National Institute of Standards and Technology. (2024). *Artificial intelligence risk management framework: Generative artificial intelligence profile*. U.S. Department of Commerce. https://nvlpubs.nist.gov/nistpubs/ai/NIST.AI.600-1.pdf

National Research Council (NRC). (1979). *Privacy and confidentiality as factors in survey nonresponse*. The National Academies Press. https://doi.org/10.17226/19845

NRC. (1984). *Cognitive aspects of survey methodology: Build a bridge between disciplines*. The National Academies Press. https://doi.org/10.17226/930

NRC. (1992). *Principles and practices for a federal statistical agency: First edition*. The National Academies Press. https://doi.org/10.17226/9043

NRC. (2001). *Principles and practices for a federal statistical agency: Second edition*. The National Academies Press. https://doi.org/10.17226/10057

NRC. (2005). *Principles and practices for a federal statistical agency: Third edition*. The National Academies Press. https://doi.org/10.17226/11252

NRC. (2009). *Principles and practices for a federal statistical agency: Fourth edition*. The National Academies Press. https://doi.org/10.17226/12564

NRC. (1993a). *Private lives and public policies: Confidentiality and accessibility of government statistics*. The National Academies Press. https://doi.org/10.17226/2122

NRC. (1993b). *The future of the Survey of Income and Program Participation*. The National Academies Press. https://doi.org/10.17226/2072

NRC. (1995). *Measuring poverty—a new approach*. The National Academies Press. https://doi.org/10.17226/4759

NRC. (1997a). *The Bureau of Transportation Statistics: Priorities for the future*. The National Academies Press. https://doi.org/10.17226/5809

NRC. (1997b). *Assessing policies for retirement income: Needs for data, research, and models*. The National Academies Press. https://doi.org/10.17226/5420

NRC. (1999). *Sowing seeds of change: Informing public policy in the economic research service of USDA*. The National Academies Press. https://doi.org/10.17226/6320

NRC. (2000a). *Small-area income and poverty estimates: Priorities for 2000 and beyond*. The National Academies Press. https://doi.org/10.17226/9957

NRC. (2000b). *Small-area estimates of school-age children in poverty: Evaluation of current methodology*. The National Academies Press. https://doi.org/10.17226/6427

NRC. (2003a). *Survey automation: Report and workshop proceedings*. The National Academies Press. https://doi.org/10.17226/10695

NRC. (2003b). *Protecting participants and facilitating social and behavioral sciences research*. The National Academies Press. https://doi.org/10.17226/10695

NRC. (2004a). *Measuring research and development expenditures in the U.S. economy*. The National Academies Press. https://doi.org/10.17226/11111

NRC. (2004b). *Climate data records from environmental satellites*. The National Academies Press. https://doi.org/10.17226/10944

NRC. (2004c). *Reengineering the 2010 Census: Risks and challenges*. The National Academies Press. https://doi.org/10.17226/10959

NRC. (2004d). *The 2000 Census: Counting under adversity*. The National Academies Press. https://doi.org/10.17226/10907

NRC. (2004e). *Eliminating health disparities: Measurement and data needs*. The National Academies Press. https://doi.org/10.17226/10979

NRC. (2005). *Expanding access to research data: Reconciling risks and opportunities.* The National Academies Press. https://doi.org/10.17226/11434

NRC. (2006a). *Improving business statistics through interagency data sharing: Summary of a workshop.* The National Academies Press. https://doi.org/10.17226/11738

NRC. (2006b). *Once, only once, and in the right place: Residence rules in the decennial census.* The National Academies Press. https://doi.org/10.17226/11727

NRC. (2007a). *Using the American Community Survey: Benefits and challenges.* The National Academies Press. https://doi.org/10.17226/11901

NRC. (2007b). *Understanding business dynamics: An integrated data system for America's future.* The National Academies Press. https://doi.org/10.17226/11844

NRC. (2008a). *Rebuilding the research capacity at HUD.* The National Academies Press. https://doi.org/10.17226/12468

NRC. (2008b). *Protecting individual privacy in the struggle against terrorists—A framework for program assessment.* The National Academies Press. https://doi.org/10.17226/12452

NRC. (2008c). *Using the American Community Survey for the National Science Foundation's Science and Engineering Workforce Statistics program.* The National Academies Press. https://doi.org/10.17226/12244

NRC. (2009a). *Reengineering the Survey of Income and Program Participation.* The National Academies Press. https://doi.org/10.17226/12715

NRC. (2009b). *Ensuring the quality, credibility, and relevance of U.S. justice statistics.* The National Academies Press. https://doi.org/10.17226/12671

NRC. (2009c). *America's energy future: Electricity from renewable resources: Status, prospects, and impediments.* The National Academies Press. https://doi.org/10.17226/12619

NRC. (2010a). *National security implications of climate change for U.S. naval forces: Letter report.* The National Academies Press. https://doi.org/10.17226/12782

NRC. (2010b). *Accounting for health and health care: Approaches to measuring the sources and costs of their improvement.* The National Academies Press. https://doi.org/10.17226/12938

NRC. (2010c). *Limiting the magnitude of future climate change.* The National Academies Press. https://doi.org/10.17226/12785

NRC. (2012a). *Medical care economic risk: Measuring financial vulnerability from spending on medical care.* The National Academies Press. https://doi.org/10.17226/13525

NRC. (2012b). *Effective tracking of building energy use: Improving the commercial buildings and residential energy consumption surveys.* The National Academies Press. https://doi.org/10.17226/13360

NRC. (2013a). *Review of the research program of the U.S. DRIVE Partnership: Fourth report.* The National Academies Press. https://doi.org/10.17226/21725

NRC. (2013b). *Nonresponse in social science surveys: A research agenda.* The National Academies Press. https://doi.org/10.17226/18293

NRC. (2013c). *Benefits, burdens, and prospects of the American Community Survey: Summary of a workshop.* The National Academies Press. https://doi.org/10.17226/18259

NRC. (2014). *Proposed revisions to the common rule for the protection of human subjects in the behavioral and social sciences.* The National Academies Press. https://doi.org/doi:10.17226/18614

National Science and Technology Council. (2023). *A framework for federal scientific integrity policy and practice.* Scientific Integrity Framework Interagency Working Group. https://www.whitehouse.gov/wp-content/uploads/2023/01/01-2023-Framework-for-Federal-Scientific-Integrity-Policy-and-Practice.pdf

Nelson, A. (2022). *Memorandum for the heads of executive departments and agencies: Ensuring free, immediate, and equitable access to federally funded research.* U.S. Office of Science and Technology Policy. https://www.whitehouse.gov/wp-content/uploads/2022/08/08-2022-OSTP-Public-access-Memo.pdf

Norwood, J. L. (1975). Should those who produce statistics analyze them? How far should analysis go? An American view. *Bulletin of the International Statistical Institute [Proceedings of the 40th Session]*, 46, 420–432. https://www.isi-web.org/

Norwood, J. L. (1995). *Organizing to count: Change in the federal statistical system.* The Urban Institute Press. https://catalog.princeton.edu/catalog/SCSB-3263340

Norwood, J. L. (2016). Politics and federal statistics. *Statistics and Public Policy*, 3(1), 1–8. http://dx.doi.org/10.1080/2330443X.2016.1241061

Office of Management and Budget (OMB). (1978). *Statistical policy directive 14: Official poverty measure.* https://www.census.gov/topics/income-poverty/poverty/about/history-of-the-poverty-measure/omb-stat-policy-14.html

OMB. (1985). *Statistical policy directive 3: Compilation, release, and evaluation of principal federal economic indicators.* 50 F.R. 38932. https://www.whitehouse.gov/wp-content/uploads/legacy_drupal_files/omb/assets/OMB/inforeg/statpolicy/dir_3_fr_09251985.pdf

OMB. (1997a). *Statistical policy directive 15: Standards for maintaining, collecting, and presenting federal data on race and ethnicity.* 62 F.R. 58782. https://www.gpo.gov/fdsys/pkg/FR-1997-10-30/pdf/97-28653.pdf

OMB. (1997b). *Order providing for the confidentiality of statistical information.* 62 F.R. 35044. https://www.govinfo.gov/content/pkg/FR-1997-06-27/pdf/FR-1997-06-27.pdf

OMB. (2002). *Guidelines for ensuring and maximizing the quality, objectivity, utility, and integrity of information disseminated by federal agencies; Republication.* 67 F.R. 8452. https://www.federalregister.gov/documents/2002/02/22/R2-59/guidelines-for-ensuring-and-maximizing-the-quality-objectivity-utility-and-integrity-of-information

OMB. (2005). *M-05-03: Final information quality bulletin for peer review.* https://www.whitehouse.gov/wp-content/uploads/legacy_drupal_files/omb/memoranda/2005/m05-03.pdf

OMB. (2006). *Statistical policy directive 2: Standards and guidelines for statistical surveys.* https://www.whitehouse.gov/wp-content/uploads/2021/04/standards_stat_surveys.pdf

OMB. (2007). *Implementation guidance for Title V of the E-Government Act, Confidential Information Protection and Statistical Efficiency Act of 2002 (CIPSEA).* 72 F.R. 33362. https://www.federalregister.gov/documents/2007/06/15/E7-11542/implementation-guidance-for-title-v-of-the-e-government-act-confidential-information-protection-and

OMB. (2008). *Statistical policy directive 4: Release and dissemination of statistical products produced by federal statistical agencies.* 73 F.R. 12622. https://www.gpo.gov/fdsys/pkg/FR-2008-03-07/pdf/E8-4570.pdf

OMB. (2013a). *Department of Health and Human Services: Centers for Disease Control and Prevention: Statement of organization, functions, and delegations of authority.* 78 F.R. 70049. https://www.govinfo.gov/content/pkg/FR-2013-11-22/pdf/2013-27088.pdf

OMB. (2013b). *M-13-13: Open data policy—managing information as an asset.* https://www.whitehouse.gov/wp-content/uploads/legacy_drupal_files/omb/memoranda/2013/m-13-13.pdf

OMB. (2014a). *M-14-06: Guidance for providing and using administrative data for statistical purposes.* https://obamawhitehouse.archives.gov/sites/default/files/omb/memoranda/2014/m-14-06.pdf

OMB. (2014b). *Statistical policy directive 1: Fundamental responsibilities of federal statistical agencies and recognized statistical units.* 79 F.R. 71610. https://www.federalregister.gov/documents/2014/12/02/2014-28326/statistical-policy-directive-no-1-fundamental-responsibilities-of-federal-statistical-agencies-and

OMB. (2015). *M-15-15: Guidance on improving statisical activities through interagency collaboration.* https://www.whitehouse.gov/wp-content/uploads/legacy_drupal_files/omb/memoranda/2015/m-15-15.pdf

OMB. (2016a). *Standards for maintaining, collecting, and presenting federal data on race and ethnicity.* 81 F.R. 67398. https://www.federalregister.gov/documents/2016/09/30/2016-23672/standards-for-maintaining-collecting-and-presenting-federal-data-on-race-and-ethnicity

OMB. (2016b). *M-16-21: Federal source code policy: Achieving efficiency, transparency, and innovation through reusable and open source software.* https://www.whitehouse.gov/wp-content/uploads/legacy_drupal_files/omb/memoranda/2016/m_16_21.pdf

OMB. (2016c). *Statistical policy directive 2 (addendum): Standards and guidelines for cognitive interviews.* 81 F.R. 70586. https://www.federalregister.gov/documents/2016/10/12/2016-24607/statistical-policy-directive-no-2-standards-and-guidelines-for-statistical-surveys-addendum

OMB. (2016d). *Guidance on agency survey and statistical information collections—questions and answers when designing surveys for information collections.* https://www.whitehouse.gov/wp-content/uploads/legacy_drupal_files/omb/assets/OMB/inforeg/pmc_survey_guidance_2006.pdf

OMB. (2016e). *Statistical policy directive 4 (addendum): Release and dissemination of statistical products produced by federal statistical agencies and recognized statistical units [adding section 10, performance review].* 81 F.R. 71538. https://www.federalregister.gov/d/2016-25049

OMB. (2016f). *Statistical policy directive no. 2: Standards and guidelines for statistical surveys.* 81 F.R. 70586. https://www.federalregister.gov/d/2016-24607

OMB. (2017a). *Statistical policy directive 10: Standard occupational classification.* 82 F.R. 56271. https://www.gpo.gov/fdsys/pkg/FR-2017-11-28/pdf/2017-25622.pdf

OMB. (2017b). *Proposals from the Federal Interagency Working Group for revision of the standards for maintaining, collecting, and presenting federal data on race and ethnicity.* 82 F.R. 12242. https://www.federalregister.gov/documents/2017/03/01/2017-03973/proposals-from-the-federal-interagency-working-group-for-revision-of-the-standards-for-maintaining

OMB. (2019a). *M-19-15: Improving implementation of the Information Quality Act.* https://www.whitehouse.gov/wp-content/uploads/2019/06/M-19-18.pdf

OMB. (2019b). *M-19-23: Phase 1 implementation of the Foundations for Evidence-Based Policymaking Act of 2018: Learning agendas, personnel, and planning guidance.* https://www.whitehouse.gov/wp-content/uploads/2019/07/m-19-23.pdf

OMB. (2019c). *M-19-18: Federal data strategy—a framework for consistency.* https://www.whitehouse.gov/wp-content/uploads/2019/06/M-19-18.pdf

OMB. (2020a). *M-20-12: Phase 4 implementation of the Foundations for Evidence-Based Policymaking Act of 2018: Program evaluation standards and practices.* https://www.whitehouse.gov/wp-content/uploads/2020/03/M-20-12.pdf

OMB. (2020b). *M-20-12 Phase 4 implementation of the Foundations for Evidence-Based Policymaking Act of 2018: Program evaluation standards and practices.* https://www.whitehouse.gov/wp-content/uploads/2020/03/M-20-12.pdf

OMB. (2021a). *Federal data strategy.* https://strategy.data.gov/assets/docs/2021-Federal-Data-Strategy-Action-Plan.pdf

OMB. (2021b). *2020 standards for delineating core based statistical areas.* 86 F.R. 37770. https://www.federalregister.gov/documents/2021/07/16/2021-15159/2020-standards-for-delineating-core-based-statistical-areas

OMB. (2022a). *M-23-04: Establishment of standard application process requirements on recognized statistical agencies and units.* https://www.whitehouse.gov/wp-content/uploads/2022/12/M-23-04.pdf

OMB. (2022b). *North American Industry Classification System—revision for 2022; Update of Statistical Policy Directive No. 8, North American industry classification system: Classification of establishments; And elimination of Statistical Policy Directive No. 9, standard industrial classification of enterprises.* 86 F.R. 72277. https://www.govinfo.gov/content/pkg/FR-2021-12-21/pdf/2021-27536.pdf

OMB. (2023a). *OMB Bulletin No. 23-01: Revised delineations of metropolitan statistical areas, micropolitan statistical areas, and combined statistical areas, and guidance on uses of the delineations of these areas.* https://www.whitehouse.gov/wp-content/uploads/2023/07/OMB-Bulletin-23-01.pdf

OMB. (2023b). *Statistical programs of the United States government: Fiscal years 2021/2022.* https://www.whitehouse.gov/wp-content/uploads/2024/02/statistical-programs-20212022.pdf

OMB. (2023c). *National strategy to develop statistics for environmental economic decisions.* https://www.whitehouse.gov/wp-content/uploads/2023/01/Natural-Capital-Accounting-Strategy-final.pdf

OMB. (2023d). *Notice of proposed rulemaking on the fundamental responsibilities of recognized statistical agencies and units.* 88 F.R. 56708. https://www.federalregister.gov/documents/2023/08/18/2023-17664/fundamental-responsibilities-of-recognized-statistical-agencies-and-units

OMB. (2023e). *Initial proposals for updating OMB's race and ethnicity statistical standards.* 88 F.R. 5375. https://www.govinfo.gov/content/pkg/FR-2023-01-27/pdf/2023-01635.pdf

OMB. (2023f). *Improving implementation of the Information Quality Act: Frequently asked questions.* https://www.whitehouse.gov/wp-content/uploads/2023/12/FAQs-Implemention-of-the-Information-Quality-Act-final.pdf

OMB. (2024a). *Analytical perspectives.* https://www.whitehouse.gov/wp-content/uploads/2024/03/ap_10_statistics_fy2025.pdf

OMB. (2024b). *Fundamental responsibilities of recognized statistical agencies and units.* 89 F.R. 82453. https://www.federalregister.gov/documents/2024/10/11/2024-23536/fundamental-responsibilities-of-recognized-statistical-agencies-and-units

OMB. (2024c). *Leveraging federal statistics to strengthen evidence-based decision-making.* https://www.whitehouse.gov/wp-content/uploads/2024/03/ap_10_statistics_fy2025.pdf

OMB. (2024d). *Statistical policy directive 3: Compilation, release, and evaluation of principal federal economic indicators.* 89 F.R. 11873. https://www.govinfo.gov/content/pkg/FR-2024-02-15/pdf/2024-02972.pdf

OMB. (2024e). *Statistical officials: Highlights and achievements, 2023.* https://www.whitehouse.gov/wp-content/uploads/2024/03/ap_10_supp_fy2025.pdf

OMB. (2024f). *M-24-10: Advancing governance, innovation, and risk management for agency use of artificial intelligence.* https://www.whitehouse.gov/wp-content/uploads/2024/03/M-24-10-Advancing-Governance-Innovation-and-Risk-Management-for-Agency-Use-of-Artificial-Intelligence.pdf

Office of Science and Technology Policy (OSTP). (2010). *Scientific integrity.* https://obamawhitehouse.archives.gov/sites/default/files/microsites/ostp/scientific-integrity-memo-12172010.pdf

OSTP. (2013). *Increasing access to the results of federally funded scientific research.* https://obamawhitehouse.archives.gov/sites/default/files/microsites/ostp/ostp_public_access_memo_2013.pdf

OSTP. (2022). *Blueprint for an AI bill of rights.* https://www.whitehouse.gov/wp-content/uploads/2022/10/Blueprint-for-an-AI-Bill-of-Rights.pdf

OSTP. (2023a). *Strengthening and democratizing the U.S. artificial intelligence innovation ecosystem: An implementation plan for a national artificial intelligence research resource.* https://www.ai.gov/wp-content/uploads/2023/01/NAIRR-TF-Final-Report-2023.pdf

OSTP. (2023b). *Policy memorandum on scientific integrity.* https://www.whitehouse.gov/wp-content/uploads/2023/06/OSTP-SCIENTIFIC-INTEGRITY-POLICY.pdf

OSTP. (2024g). *Revisions to OMB's statistical policy directive no. 15: Standards for maintaining, collecting, and presenting federal data on race and ethnicity.* 89 F.R. 22182. https://www.federalregister.gov/documents/2024/03/29/2024-06469/revisions-to-ombs-statistical-policy-directive-no-15-standards-for-maintaining-collecting-and

Paperwork Reduction Act, Pub. Law 96-511, 94 Stat. 2812. (1980). https://uscode.house.gov/statutes/pl/96/511.pdf

Paperwork Reduction Act, Pub. Law 99-500, § 146. (1986).

Paperwork Reduction Act, Pub. Law 99-591, § 101 (m) 100 Stat. 3341-308, 3341-335 (1986). https://www.govinfo.gov/content/pkg/STATUTE-100/pdf/STATUTE-100-Pg3341.pdf

Paperwork Reduction Act, Pub. L. No. 104-13, 109 Stat. 163. (1995). https://www.govinfo.gov/content/pkg/PLAW-104publ13/html/PLAW-104publ13.htm

Park, J., & Tractenberg, R. E. (2023). *How do ASA ethical guidelines support U.S. guidelines for official statistics?* Ethics International Press. https://arxiv.org/pdf/2309.07180

Prell, M., Chapman, C., Adeshiyan, S., Fixler, D., Garin, T., Mirel, L. B., & Phipps, P. (2019). *Transparent reporting for integrated data quality: Practices of seven federal statistical agencies.* Federal Committee on Statistical Methodology. https://nces.ed.gov/FCSM/pdf/Transparent_Reporting_FCSM_19.01.pdf#:~:text=This%20report%20provides%20results%20from%20the

Privacy Act of 1974, Pub. L. No. 93-579, 88 Stat. 1896. (1974). https://www.govinfo.gov/content/pkg/STATUTE-88/pdf/STATUTE-88-Pg1896.pdf

Pub. Law 104-191, 110 Stat. 1936. (1996). https://www.congress.gov/104/plaws/publ191/PLAW-104publ191.pdf

Pub. Law 114-94, 130 Stat. 538. (2016). https://www.congress.gov/114/plaws/publ185/PLAW-114publ185.pdf

Reamer, A. D. (2014). *Stumbling into the Great Recession: How and why GDP estimates kept economists and policymakers in the dark.* The George Washington Institute of Public Policy. https://gwipp.gwu.edu/sites/g/files/zaxdzs2181/f/downloads/Reamer_GDP_Research_Note_04-25-14.pdf

Royal Statistical Society. (2014). *Code of conduct.* https://rss.org.uk/RSS/media/File-library/About/2019/RSS-Code-of-Conduct-2014.pdf

Ryten, J. (1990). Statistical organization criteria for inter-country comparisons and their application to Canada. *Journal of Official Statistics,* 6(3), 319–332. https://www.proquest.com/docview/1266811482?pq-origsite=gscholar&fromopenview=true&sourcetype=Scholarly%20Journals

Science, State, Justice, Commerce, and Related Agencies Appropriations Act, Pub. L. No. 109-108, 119 Stat. 2308. (2005). https://www.congress.gov/109/plaws/publ108/PLAW-109publ108.pdf

Temporary Assignments Under the Intergovernmental Personnel Act, 5 C.F.R. Part 334. (2024). https://www.ecfr.gov/current/title-5/chapter-I/subchapter-B/part-334

Title 42—The public health and welfare, 5137 § 1885d. (2024). https://statecodesfiles.justia.com/us/2022/title-42/chapter-16/sec-1885d/sec-1885d.pdf?ts=1722330151

Title 49—Transportation, 350 § 49 § 6302. (2024). https://statecodesfiles.justia.com/us/2018/title-49/subtitle-iii/chapter-63/sec-6302/sec-6302.pdf?ts=1588215779

Tractenberg, R. E. (2020). *Concordance of professional ethical practice standards for the domain of data science: A white paper.* Open Archive of the Social Sciences (SocArXiv). https://osf.io/preprints/socarxiv/p7rj2

Tractenberg, R. E. (2022a). *Ethical reasoning for a data-centered world.* Ethics International Press. https://ethicalreasoning.org/books/

Tractenberg, R. E. (2022b). *Ethical practice in statistics and data science*. Ethics International Press. https://ethicalreasoning.org/books/

Tractenberg, R. E., & Park, J. (2023). *How does international guidance for statistical practice align with the ASA Ethical Guidelines?* Ethics International Press. https://arxiv.org/pdf/2309.08713

Triplett, J. (1991). The federal statistical system's response to emerging data needs. *Journal of Economic and Social Measurement, 17*(3-4), 155–201. https://content.iospress.com/articles/journal-of-economic-and-social-measurement/jem17-3-4-03#:~:text=The%20Federal%20Statistical%20System's%20Response%20to

United Nations Statistical Commission. (2014). *Fundamental principles of official statistics*. United Nations. https://unstats.un.org/unsd/dnss/gp/FP-New-E.pdf

Voting Rights Act of 1965, Pub. L. No. 89-110. (1965). https://www.govinfo.gov/content/pkg/STATUTE-79/pdf/STATUTE-79-Pg437.pdf

Young, L. J. (2019). Agricultural crop forecasting for large geographical areas. *Annual Review of Statistics and Its Application, 6*, 173–196. https://doi.org/10.1146/annurev-statistics-030718-105002

Young, L. J., & Chen, L. (2022). Using small area estimation to produce official statistics. *Stats, 5*(3), 881–897. https://www.mdpi.com/2571-905X/5/3/51

Appendixes A, B, and C are available online at
https://nap.nationalacademies.org/catalog/27934.

Appendix D

Biographical Sketches

KATHARINE G. ABRAHAM is a distinguished professor of economics and survey methodology at the University of Maryland, College Park. She formerly served as commissioner of the U.S. Bureau of Labor Statistics, as a member of the President's Council of Economic Advisers, and, most recently, as chair of the U.S. Commission on Evidence-based Policymaking. Abraham's published research includes papers on the contingent workforce, the work and retirement decisions of older Americans, unemployment and job vacancies, and the measurement of economic activity. She is an elected member of the American Academy of Arts and Sciences, a distinguished fellow of the American Economic Association, and a fellow of both the American Statistical Association and the Society of Labor Economists. Abraham has a Ph.D. in economics from Harvard University. She is an elected member of the National Academy of Sciences, is currently serving as Chair of the Committee on National Statistics, and has served on several consensus panels for the National Academies of Sciences, Engineering, and Medicine.

MICK P. COUPER is a research professor at the Survey Research Center in the Institute for Social Research at the University of Michigan. His current research interests include survey non-response, design and implementation of survey data collection, effects of technology on the survey process, and computer-assisted interviewing, including both interviewer-administered (CATI and CAPI) and self-administered (web, audio-CASI, IVR) surveys. Many of Couper's current projects focus on the design of web, smartphone, and mixed-mode surveys. He has an M.Soc.Sc. in sociology from the University of Cape Town, South Africa, an M.A. in applied social research from the University of Michigan, and a Ph.D. in sociology from Rhodes University in South Africa. Couper is a current member of the National Academies' Committee on National Statistics and has previously served on their Panel on Redesigning the BLS Consumer Expenditures Surveys, Panel on the Research on Future Census Methods, and the Oversight Committee for the Workshop on Survey Automation.

WILLIAM A. DARITY, JR. is the Samuel DuBois Cook Professor of Public Policy, African and African American Studies, Economics, and Business and the director of the Samuel DuBois Cook Center on Social Equity at Duke University. He has served as chair of the Department of African and African American Studies and was the founding director of the Research Network on Racial and Ethnic Inequality at Duke. Darity's research focuses on inequality by race, class and ethnicity, stratification economics, skin shade and labor market outcomes, the economics of reparations, the Atlantic slave trade and the Industrial Revolution, and the social psychological effects of exposure to unemployment. He was a visiting scholar at the Russell Sage Foundation and a fellow at the Center for Advanced Study in the Behavioral Sciences at Stanford. Darity is also a past president of the National Economic Association, the Southern Economic Association, and the Association of Black Sociologists. He was named as a Distinguished Fellow of the American Economic Association, a W.E.B. Du Bois Fellow of the American Academy of Political and Social Sciences, and a Distinguished Fellow of the Southern Economic Association. Darity received his Ph.D. in economics from the Massachusetts Institute of Technology.

DIANA FARRELL is an independent director and trustee of various organizations, including the Urban Institute, the National Bureau of Economic Research, the Institute for Applied and Practical Mathematics, a National Science Foundation Center at the University of California Los Angeles, and until recently, eBay. She was the founding president and chief executive officer of the JPMorgan Chase Institute, where she created a legacy of producing and publishing unique data analyses and insights leveraging the bank's administrative transactions data. Previously, Farrell was a senior partner at McKinsey & Company, where she served on the Partner's Evaluation committee, and was the founder and Global Head of the McKinsey Center for Government as well as the Global Head of the McKinsey Global Institute. At various points in her McKinsey career, she was a leader in the public sector, the financial institutions sector, and the strategy practice. Additionally, Farrell served in the White House for over 2 years as deputy director of the National Economic Council and deputy assistant to the president on Economic Policy. She is a member of the Council on Foreign Relations, the Aspen World Economy Group, the Aspen Strategy

Group, the Trilateral Commission, and the Bretton Woods Committee. Farrell holds a B.A. from Wesleyan University, where she was awarded a Distinguished Alumna award and is a trustee emeritus, and an M.B.A. from Harvard Business School. She is a member of the National Academies of Sciences' Committee on National Statistics.

ROBERT M. GOERGE is a senior fellow at the National Opinion Research Center (NORC) at the University of Chicago. He has focused for the past 35 years on improving the data and evidence consumed by policymakers, administrators, and practitioners in social programs at the federal, state and local levels, specifically income maintenance, child welfare, primary and postsecondary education, criminal and juvenile justice, and early childhood programs. He focuses on the rigorous development and collection of data that accurately and comprehensively reflects the experiences of children and families in social programs. Goerge has developed and employed record-linkage methods to combine large data sources. He has had an IPA position at Census. Goerge is also a senior fellow at the Harris School of Public Policy, where he teaches, and a senior fellow at NORC. He has a Ph.D. from the University of Chicago in social policy. Goerge is currently a member of the Committee on National Statistics.

ERICA GROSHEN is senior labor economics advisor at Cornell University School of Industrial and Labor Relations, research fellow at the Upjohn Institute for Employment Research, and a member of the Federal Economic Statistics Advisory Committee. She previously served as the 14th Commissioner of the U.S. Bureau of Labor Statistics, the principal federal agency responsible for measuring labor market activity, working conditions, and inflation. Before that Groshen was vice president in the Research and Statistics Group of the Federal Reserve Bank of New York. Her research centers on employers' roles in labor market outcomes. She co-edited "Improving Employment and Earnings in Twenty-First Century Labor Markets" (*The Russell Sage Foundation Journal of the Social Science*), co-authored *How New is the 'New Employment Contract'?* (W.E. Upjohn Institute Press), and co-edited *Structural Changes in U.S. Labor Markets: Causes and Consequences* (M.E. Sharpe, Inc.). Groshen received the Susan C. Eaton Outstanding Scholar-Practitioner Award from the Labor and Employment Relations Association and was appointed a fellow of the American Statistical Association. She holds a B.S. in mathematics and economics from the University of Wisconsin-Madison and a Ph.D. in economics from Harvard University. Groshen is a member of the Committee on National Statistics.

DANIEL HO is the William Benjamin Scott and Luna M. Scott Professor of Law at Stanford Law School, professor of political science, and professor of computer science (by courtesy) at Stanford University. He is also a senior fellow at the Stanford Institute for Economic Policy Research and the Stanford Institute for Human-Centered Artificial Intelligence, faculty fellow at the Center for Advanced Study in the Behavioral Sciences, and director of the Regulation, Evaluation, and Governance Lab (RegLab). Ho serves as an appointed member to the National Artificial Intelligence Advisory Commission and as a public member of the Administrative Conference of the United States. His scholarship focuses on administrative law, regulatory policy, and antidiscrimination law. With the RegLab, Ho's work has developed high-demonstration projects of data science and machine learning in public policy, through partnerships with a range of government agencies, including the Internal Revenue Service, the Department of the Treasury, the Environmental Protection Agency, the Santa Clara County Public Health Department, and Seattle and King County Public Health. He also clerked for Judge Stephen F. Williams on the U.S. Court of Appeals, District of Columbia Circuit. He is the recipient of numerous awards, including the John Bingham Hurlbut Award for Excellence in Teaching at Stanford Law School, the Carole Hafner Award for the best paper at the International Conference on Artificial Intelligence and Law, the Best Empirical Paper Prize from the American Law and Economics Review, and the Warren Miller prize for the best paper published in *Political Analysis*. Ho received his J.D. from Yale Law School and Ph.D. from Harvard University.

HILARY HOYNES is Professor of Economics and Public Policy and holds the Haas Distinguished Chair of Economics Disparities at the University of California, Berkeley. She also directs the Berkeley Opportunity Lab. She is an economist who works on poverty, inequality, and the social safety net. Her current research examines how access to the social safety net in early life affects children's later life health and human capital outcomes. Hoynes is a member of the National Academy of Sciences, the American Academy of Arts and Sciences, and the National Academy of Social Insurance, and a fellow of the Society of Labor Economists. She has served as co-editor of the *American Economic Review* and the *American Economic Journal: Economic Policy*. She is a member of the Committee on National Statistics and serves on California Governor Gavin Newsom's Council of Economic Advisors. Previously, she served as Vice President of the American Economic Association and she served on the National Academy of Sciences Committee on Building an Agenda to Reduce the Number of Children in Poverty by Half in 10 Years, the State of California Task Force on Lifting Children and Families out of Poverty, and the Federal Commission on Evidence-Based Policy Making. She received the Carolyn Shaw Bell Award from the Committee on the Status of the Economics Profession of the American Economic Association. Hoynes received her Ph.D. in economics from Stanford University and her undergraduate degree in economics and mathematics from Colby College.

H. V. JAGADISH is Edgar F. Codd Distinguished University Professor and Bernard A. Galler Collegiate Professor of Electrical Engineering and Computer Science at the University of Michigan in Ann Arbor, and director of the Michigan Institute for Data Science. Previously, he was head of the Database Research Department at AT&T Labs. Jagadish is well known for his broad-ranging research on information management, and has approximately 200 major papers and 37 patents, with an H-index of 94. He is a fellow of the Association for Computing Machinery (ACM) and of the American Association for the Advancement of Science. Jagadish has served on the board of the Computing Research Association. He has been an associate editor for the *ACM Transactions on Database Systems*, program chair of the ACM Special Interest Group on Management of Data (SIGMOD) annual conference, program chair of the International Society for Computational Biology conference, a trustee of the Very Large DataBase (VLDB) foundation, founding editor-in-chief of the *Proceedings of the VLDB Endowment*, and program chair of the VLDB Conference. He is editor of the Morgan & Claypool "Synthesis" Lecture Series on Data Management. Jagadish won the David E. Liddle Research Excellence Award (at the University of Michigan), the ACM SIGMOD Contributions Award, and the Distinguished Faculty Achievement Award (at the University of Michigan). His massive open online course on data science ethics is available on EdX, Coursera, and Futurelearn.

DANIEL KIFER is a professor in the Department of Computer Science and Engineering at Penn State University. He works on machine learning and security, with particular emphasis on privacy technology. Kifer served as a technical lead for the Census Bureau's disclosure avoidance system and also consults for Facebook on privacy technology. His work has been recognized with distinctions such as the Association for Computing Machinery (ACM) Special Interest Group on Management of Data test of time award, the Institute of Electrical and Electronics Engineers International Conference on Data Engineering influential paper award, the ACM Computing Classification System outstanding paper award, and the Caspar Bowden Privacy Enhancing Technologies Award. Kifer is currently a member of the Committee on National Statistics.

SHARON L. LOHR is a professor emerita at Arizona State University, where she was Dean's Distinguished Professor of Statistics. When she was a vice president at Westat, she developed survey designs and statistical analysis methods for use in transportation, public health, crime measurement, and education. Lohr's research interests include sample surveys, design of experiments, hierarchical models, and combining multiple sources of data. She is the author of numerous research articles as well as the books *Sampling: Design and Analysis* and *Measuring Crime: Behind the Statistics*. She is an elected fellow of the American Statistical Association, an elected member of the International Statistical Institute, and the inaugural recipient of the Gertrude M. Cox Statistics Award for contributions to the practice of statistics. Lohr's invited presentations include the Morris Hansen, Deming, and Waksberg lectures. She earned her B.S. degree in mathematics from Calvin College, and her Ph.D. in statistics from the University of Wisconsin-Madison. Lohr currently serves on the Committee on National Statistics and has also served on three previous National Academies' committees, serving as chair of the Panel on the Implications of Using Multiple Data Sources for Major Survey Programs, and as a member of the Panel on Improving Federal Statistics for Policy and Social Science Research Using Multiple Data Sources and State-of-the-Art Estimation Methods and of the Panel on the Functionality and Usability of Data from the American Community Survey.

NELA RICHARDSON is ADP's chief economist and environmental, social and governance officer. She is the head of the ADP Research Institute (ADPRI). Richardson's background and expertise cross many industries, including finance, technology, housing, and labor. In response to the accelerated pace of economic change, she led the launch of a high-frequency revamp of the renowned ADP National Employment Report in collaboration with Stanford Digital Economy Lab. In addition to ongoing labor market analysis, Richardson provides insights on the dynamic shifts of the economy. Prior to her work at ADP, she was principal and investment strategist at Edward Jones. Richardson has held research positions at the Commodity Futures Trading Commission, Harvard University's Joint Center for Housing Studies, and Freddie Mac. She also worked as an adjunct finance professor at the Carey School of Business at John Hopkins University. Richardson has a bachelor's degree in mathematics, economics, and philosophy from Indiana University in Bloomington, an M.S. in economics from the University of Pennsylvania, and a Ph.D. in economics, with concentrations in financial economics, international finance, and economic development, from the University of Maryland, College Park.

C. MATTHEW SNIPP is Burnet C. and Milfred Finley Wohlford professor of sociology at Stanford University. At Stanford, he currently serves as director of the Secure Data Center, deputy director of the Institute for Research in the Social Sciences, and chair of the Native American Studies program. Snipp has written extensively on Native Americans, focusing specifically on the interaction of Native Americans and the U.S. Census. Snipp has served on the Census Bureau's Technical Advisory Committee on Racial and Ethnic Statistics and the Native American Population Advisory Committee. He is the former director of the Center for Comparative Studies of Race and Ethnicity. Prior to moving to Stanford, Snipp was associate professor and professor of rural sociology at the University of Wisconsin–Madison, where he held affiliate appointments with several other units, and assistant and associate professor of sociology at the University of Maryland. He has an M.S. and a Ph.D. in sociology from the University of Wisconsin, Madison. Snipp is a member of the National Academies' Committee on National Statistics, and he previously served as a member of the Panel to Review the 2010 Census, the Panel on Residence Rules in the Decennial Census, and the Panel on the Research on Future Census Methods, and as co-chair of the Steering Committee for a Workshop on Developing a New National Survey on Social Mobility.

ELIZABETH A. STUART is Hurley-Dorrier Professor and Chair and Bloomberg Professor of American Health in the Department of Biostatistics at the Johns Hopkins Bloomberg School of Public Health, with joint appointments in the Department of Mental Health and the Department of Health Policy and Management. She has extensive experience in methods for estimating causal effects for program and policy evaluation, particularly as applied to mental health, public policy, and education. Stuart's primary research interests include designs for estimating causal effects in non-experimental settings (such as propensity scores) and methods to assess and enhance the generalizability of randomized trials to target populations. Stuart has received research funding from the National Science Foundation, the Patient-Centered Outcomes Research Institute, the Institute of Education Sciences, the WT Grant Foundation, and the National Institutes of Health. Stuart is a fellow of the American Statistical Association (ASA) and the American Association for the Advancement of Science, and has received the mid-career award from the Health Policy Statistics Section of the ASA, the Gertrude Cox Award for applied statistics, Harvard University's Myrto Lefkopoulou Award for excellence in biostatistics, the Rod Little Lectureship from the University of Michigan's Department of Biostatistics, and the inaugural Society for Epidemiologic Research Marshall Joffe Epidemiologic Methods award. She received her Ph.D. in statistics from Harvard University. Stuart is a member of the National Academies' Committee on National Statistics and co-chair of the National Academies' Committee on Applied and Theoretical Statistics.

COMMITTEE ON NATIONAL STATISTICS

The Committee on National Statistics was established in 1972 at the National Academies of Sciences, Engineering, and Medicine to improve the statistical methods and information on which public policy decisions are based. The committee carries out studies, workshops, and other activities to foster better measures and fuller understanding of the economy, the environment, public health, crime, education, immigration, poverty, welfare, and other public policy issues. It also evaluates ongoing statistical programs and tracks the statistical policy and coordinating activities of the federal government, serving a unique role at the intersection of statistics and public policy. The committee's work is supported by a consortium of federal agencies through a National Science Foundation grant, a National Agricultural Statistics Service cooperative agreement, and several individual contracts.